TOMMY TOBIN, EDITOR

FOOD LAW

A PRACTICAL GUIDE

Cover design by Tahiti Spears/ABA Design

The materials contained herein represent the opinions of the authors and/or the editors and should not be construed to be the views or opinions of the law firms or companies with whom such persons are in partnership with, associated with, or employed by, nor of the American Bar Association or the Section of Science and Technology Law, unless adopted pursuant to the bylaws of the Association.

Nothing contained in this book is to be considered as the rendering of legal advice for specific cases, and readers are responsible for obtaining such advice from their own legal counsel. This book is intended for educational and informational purposes only.

© 2021 American Bar Association. All rights reserved.

No part of this publication may be reproduced, stored in a retrieval system, or transmitted in any form or by any means, electronic, mechanical, photocopying, recording, or otherwise, without the prior written permission of the publisher. For permission, complete the request form at www.americanbar.org/reprint or email ABA Publishing at copyright@americanbar.org.

Printed in the United States of America.

25 24 23 22   5 4 3 2

A catalog record for this book is available from the Library of Congress.

Discounts are available for books ordered in bulk. Special consideration is given to state bars, CLE programs, and other bar-related organizations. Inquire at Book Publishing, ABA Publishing, American Bar Association, 321 N. Clark Street, Chicago, Illinois 60654-7598.

www.shopABA.org

To my mother. Love you Mom.

Contents

About the Authors ... xi
Preface ... xvii
Acknowledgments ... xxi
Foreword ... xxiii

CHAPTER 1

Regulation of Food Formulation, Manufacturing, Labeling, and Advertising: A Primer, Farm to Fork ... 1
Ricardo Carvajal

Introduction ... 1
Federal Agencies That Regulate Food ... 3
    FDA ... 3
    U.S. Department of Agriculture ... 4
    Alcohol and Tobacco Tax and Trade Bureau ... 4
    U.S. Environmental Protection Agency ... 5
    Federal Trade Commission ... 5
Food Formulation ... 5
    Food Additive Approval ... 6
    "Generally Recognized as Safe" Exception ... 6
    Dietary Supplements ... 9
Food Manufacture: Ensuring Safety ... 10
Labeling and Advertising ... 18
    Mandatory Labeling Requirements ... 19
    Voluntary Labeling Requirements ... 19
Concluding Thoughts ... 22

CHAPTER 2

Food Litigation: An Emerging Field   23
Tommy Tobin

Food and Civil Litigation	23
Rise of Food Litigation	24
Types of Cases	27
Conclusion	35

CHAPTER 3

Reasoning with the Reasonable Consumer Standard in Food Litigation   37
Charles C. Sipos, Carrie Akinaka, and Tommy Tobin

Introduction	37
Doctrinal Origins	38
Rise of Food Litigation	40
What Is the Reasonable Consumer Defense and When Does It Come into Play?	42
Ongoing Debates in Reasonable Consumer Jurisprudence	45
The Reasonable Consumer Standard Is an Objective Test	46
Just How Sophisticated Is the Reasonable Consumer?	46
Courts Rely on a Number of Factors in Determining Reasonableness	47
Conclusion	52

CHAPTER 4

Food Safety   53
James F. Neale and Benjamin P. Abel

Introduction to the Regulatory Framework	53
The Major Food Regulators at the Federal Level	54
USDA	54
FDA	57
Inspections	60
Prohibited Acts	62
Allergens	65
Preventive Controls and Countermeasures	66

CHAPTER 5

Food Recalls — 73
James F. Neale and Benjamin P. Abel

Governmental Agencies Responsible for Food Product Recalls	73
Regulators Governing Food Product Recalls	74
Recall Types	75
Company-Initiated Recalls	75
Government-Initiated Recalls	76
Manufacturer's Failure or Refusal to Initiate Voluntary Recall	76
Recall Classifications	78
Class I Recalls	78
Class II Recalls	79
Class III Recalls	79
Product Withdrawals	79
Recall Plan and Team	80
Conducting a Recall	82
The Recall's Scope	84
Recall Communications	85
Status Reports	86
Regulator Site Inspection Contemporaneous with Recall	86
Admissibility of Recall in Subsequent Litigation	89
Effectiveness Checks	91
Termination	92

CHAPTER 6

Federal Nutrition Programs — 93
Roger R. Szemraj and Stewart Fried

Overview	93
SNAP	93
Child Nutrition Programs	97
How Children Qualify for Free and Reduced-Price Meals	99
What Meals May Be Served	101
Buy American Requirement	102
State Administrative Expenses	102
COVID-19 Activities	102
Summer Food Service Program	103

CACFP	103
Eligibility	103
Reimbursement Rates	104
Special Supplemental Nutrition Program for Women, Infants, and Children	105
The Food Package	106
Where WIC Benefits Can Be Redeemed	106
The Emergency Food Assistance Program	106
What Foods Are Available	107
What Operating Assistance Is Available	107
Who Is an Eligible Recipient?	107
COVID-19 Response	107
Commodity Supplemental Food Program	108
USDA Foods	108
Dietary Guidelines for Americans	109
Process	110
Importance of the Guidelines to Nutrition Programs	110
Opportunities for Involvement	110

CHAPTER 7

Practicing International Food Law: Considerations for Counsel   113

Charles F. Woodhouse

Is There Such a Thing as "International Food Law"?	113
Getting Started in International Food Law: Education	115
The Truly "International" Food Regulatory System	116
The "Work" of a U.S. "International Food Lawyer"	117
How to Train to Become an "International" Food Attorney	119
Focus Areas	120
Discussion of Packaging and Food Contact Materials	120
Allergen Labeling	121
Pesticide and Biocide Residual Levels	123
Food Labeling—International Trade	123
Food Safety Documentation	123
Recalls and Recall Management	123
Specific Issues in European Union Food Law Research	124
Conclusion	125

CHAPTER 8

Blockchain in the Food Industry   127
Darin Detwiler

Introduction	127
A "New Era of Smarter Food Safety"	128
Food Safety History	129
The "Five Pillars" of the Food System	131
Food Safety	131
Food Quality	132
Food Authenticity	132
Food Defense	133
Food Security	134
Blockchain and the Greater Technology Ecosystem	135
The Food Technology Ecosystem	135
Blockchain	137
Conclusion	140

CHAPTER 9

Food Law and the Pandemic: Securing the Food System   143
Lawrence Reichman and Tommy Tobin

COVID-19 from the Federal Agency Perspective	143
Mitigating Disruptions to the Food Supply Chain	147
Food Work Environments	150
Addressing Food Insecurity	155
What Is Ahead	157

CHAPTER 10

Intellectual Property   159
Anne W. Glazer

Introduction	159
Patent	161
The Patent Definition	161
On-Sale Bar	162
Utility Patents	163
Design Patents	163

The Effect of a Patent	164
Patent Infringement Remedies	164
Trade Secret	164
Keeping Trade Secrets Secret	165
The Trade Secret Definition	165
Misappropriation	167
Remedies	167
NDAs	168
Employees Coming and Going	169
Trademark	170
A Word about Domain Names	171
State Law, Federal Law	171
The Trademark Definition	173
Trademark Spectrum of Distinctiveness	174
Trademark Selection and Clearance	176
The Likelihood of Confusion	176
Trademark Registration	178
Trademark Infringement Remedies	179
Copyright	179
The Copyright Definition	180
Copyright Registration	180
Copyright Ownership	181
Copyright Fair Use	182
Copyright Infringement and Remedies	182
Exhaustion and First-Sale Doctrines	183
Table of Cases	*185*
Index	*191*

About the Authors

Benjamin P. Abel, McGuireWoods
Benjamin P. Abel is an Associate in McGuireWoods in its Charlottesville office. He is a seasoned litigator with extensive experience representing clients in both federal and state courtrooms. His practice focuses on representing clients in products liability and class actions, with a particular emphasis on food and beverage labeling claims. Ben earned a B.A. from the University of Virginia and his J.D. from William & Mary.
Email: BAbel@mcguirewoods.com

Carrie Akinaka, Perkins Coie LLP
Carrie Akinaka is an Associate in Perkins Coie's Los Angeles office. She is a trained persuasive writer and oral advocate—a strategic storyteller. Carrie has a varied commercial litigation practice, supporting a Fortune 5 retailer in employment and product liability actions, major real estate clients in multimillion-dollar mediations, and a multinational food company in class actions and appeals in the U.S. Court of Appeals for the Ninth Circuit.
Email: CAkinaka@perkinscoie.com

Ricardo Carvajal, Hyman, Phelps & McNamara, PC
Ricardo Carvajal is a Director at Hyman, Phelps & McNamara, PC in Washington, DC. His practice focuses on food regulatory requirements enforced by the FDA, USDA, and FTC. Mr. Carvajal advises clients ranging from start-ups to multinational companies at all points in the supply chain, including bulk ingredient manufacturers, distributors of packaged goods, and retailers. He counsels clients on the regulatory status of ingredients and finished products, and provides advice on compliance with labeling and advertising requirements. He helps clients interpret and comment on the implementation of new legislation and regulations. He also has expertise in FDA's regulation of products derived through the use of emerging technologies.
Email: RCarvajal@HPM.com

Dr. Darin Detwiler, Northeastern University
Darin Detwiler, L.P.D., M.A.Ed., is the Assistant Dean of Academic and Faculty Affairs at Northeastern University's College of Professional Studies in Boston, MA. He is also an Assistant Teaching Professor of Food Policy who serves as the lead faculty for the college's MS in Regulatory Affairs of Food and Food Industry program. He is the recipient of the College of Professional Studies' 2016 Excellence in Teaching Award. Dr. Detwiler is an internationally recognized, respected, and well-connected food policy and technology expert with 25+ years' experience in shaping federal food policy, consulting with corporations, and contributing thought leadership to industry events and publications. He is a recipient of the 2018 Distinguished Service Award from the International Association for Food Protection. Detwiler is regarded as "one of America's leading food safety advocates." In 2004, the Secretary of Agriculture appointed Detwiler to two terms on the USDA's National Advisory Committee. He later advised the FDA as the Senior Policy Coordinator for a leading national food safety advocacy organization.

Email: d.detwiler@northeastern.edu

Stewart Fried, OFW Law
Stewart Fried is a Principal at OFW Law in Washington, DC. He has broad expertise in all SNAP and WIC-related matters and represents associations, supermarkets, convenience stores, and other SNAP-eligible retailers on a wide variety of issues before FNS, Congress, and in federal courts across the country, including providing statutory and regulatory advice to and representation of retailers on issues ranging from authorization to administrative and judicial review proceedings from denials and withdrawals of SNAP authorization, trafficking in SNAP benefits. Stewart also represents food manufacturers, co-packers, retailers, and associations on food labeling, employment, corporate governance, and antitrust issues.

Email: sfried@ofwlaw.com

Anne Glazer, Stoel Rives
Anne Glazer is a Partner in Stoel Rives' Portland, Oregon office. Anne handles the broad range of product development, branding, marketing, sales, and distribution issues for food, beverage, and cosmetics clients. She regularly works with clients of all sizes, from startups to Fortune 500 companies. Anne works with clients to manage their legal costs through strategy development, prioritization, cooperation with in-house staff, creative fee arrangements, managing to budgets, and client staff training. Anne is a past co-chair of the firm's Agribusiness, Food and Beverage Industry Group.
Email: Anne.Glazer@stoel.com

James F. Neale, McGuireWoods
James F. Neale is a partner in McGuireWoods' Charlottesville office, whose varied practice focuses on high-exposure cases in courts across the country. He has substantial mass tort and class action litigation experience and currently serves as co-chair of the firm's Food & Beverage Industry Team. He holds a B.A. and J.D. from the University of Virginia.
Email: JNeale@mcguirewoods.com

Larry Reichman, Perkins Coie LLP
Larry Reichman, a Partner in Perkins Coie's Portland office, guides clients within such regulated industries as food and beverage, dietary supplements, telecommunications and energy in major initiatives and new market opportunities by resolving regulatory challenges, project impediments and related litigation. From guidance on label requirements for food and beverage products, including the unique mandates governing online sales and organic products, to advising on a major educational institution's creation of a statewide broadband network, Larry introduces business practicality into his regulatory compliance counsel to global manufacturers, corporations, public utilities, and entrepreneurs.
Email: LReichman@perkinscoie.com

Michael T. Roberts, UCLA School of Law
Michael T. Roberts is the founding Executive Director of the Resnick Center for Food Law and Policy at UCLA School of Law, where he is also a Professor from Practice. Roberts is a thought leader in a broad range of legal and policy issues from farm to fork in local, national, and global food supply systems. He taught the first food law and policy course in the United States in 2004 and served as the leading force in the development in 2005 of the first scholarly journal—*Journal of Food Law and Policy*—devoted exclusively to the field. Roberts has authored *Food Law in the United States* (Cambridge University Press 2016) and is actively involved in the global development of food law and policy.
Email: MRoberts@law.ucla.edu

Roger R. Szemraj, OFW Law
Roger R. Szemraj is a Principal at OFW Law in Washington, DC. He advises associations and individual clients on agricultural appropriations; Farm Bills; food assistance programs; international food aid and development programs; food safety issues; trade issues; and research programs. He has more than 30 years of experience on Capitol Hill, concentrating on these matters, having served as chief of staff to three House members and associate staff to the House Appropriations Committee.
Email: rszemraj@ofwlaw.com

Tommy Tobin, Perkins Coie LLP
Thomas Tobin is an Associate in Perkins Coie's Seattle office. His practice focuses on complex commercial litigation and class action matters involving statutory, constitutional, and regulatory issues in a range of industries, including food and beverage, consumer packaged goods, and cannabis. In the food and beverage sector, Tommy has experience defending false advertising claims and consumer protection claims for well-known international corporations. He regularly writes articles on food law and policy issues and is chair of the American Bar Association's Food, Cosmetics, and Nutraceuticals Committee.
Email: TTobin@perkinscoie.com

Charles Sipos, Perkins Coie LLP

Charles Sipos is an experienced class action litigator, having successfully represented clients in class action lawsuits in the technology, food and beverage, gaming, and pharmaceutical industries over the past 16 years. He has appeared and argued on behalf of defendants in class actions in courts nationwide, including in California, Colorado, the District of Columbia, Florida, Illinois, New Jersey, New York, Oregon, Washington, and the U.S. Courts of Appeals for the Ninth and Seventh Circuits. He has presented arguments resulting in dismissals and summary judgment based on federal preemption, lack of standing, mootness, primary jurisdiction, failure to allege damages, the "reasonable consumer" standard and related defenses.

Email: CSipos@perkinscoie.com

Charles F. Woodhouse, Woodhouse Shanahan PA

Charles F. Woodhouse is a partner at the Washington, DC, and Florida offices of Woodhouse Shanahan PA. He is dedicated to the details of food labeling, food safety, and food science. He holds a B.A. from Dartmouth, an M.B.A. from Wharton, a J.D. from Rutgers, and Master's degrees in Food Safety and Packaging from Michigan State University. He also is currently a graduate student in Soil and Water Science at the University of Florida's Institute of Food and Agricultural Sciences.

Email: cfw@regulatory-food-science.com

Preface

Food law is a hot topic in today's legal practice. For one thing, food is a highly regulated product with fragmented oversight on the federal and state levels. Producers and retailers face an array of compliance obligations, especially as they operate in an increasingly interconnected, globalized world. Litigation risks are on the rise for food companies, with a food litigation practice area emerging in both the plaintiff and defense bars. Meanwhile, producers are adapting to developments in technology, such as blockchain, and so, too, are regulators, who have announced a "New Era of Smarter Food Safety."[1] Indeed, the Food Safety Modernization Act (FSMA) ushered in a sea change as to the federal regulation of the food system. It is hard to believe it has only been a little more than a decade since FSMA's effective date, with its sweeping changes and new requirements. Meanwhile, consumers are facing new realities, not least of which are due to the ongoing coronavirus pandemic, which has drastically altered consumer buying patterns. The economic consequences of the pandemic have left many families with gaping holes in their food budgets due to furloughs or layoffs, in addition to the terrible human costs associated with the illness.

The study of food, from both legal and policy perspectives, has become a "flourishing" field with leading scholars proclaiming it is now a "a permanent fixture within the American legal academy."[2] Within law schools, food law and policy is now "firmly rooted as a growing and thriving legal field" and these "courses—and the faculty who teach them—are now the norm at many of America's top schools, with more than a dozen schools offering two or more such courses each year."[3]

Even while the number of law school courses on food law proliferates to meet growing student demand, substantial disagreement remains over just what constitutes *food law*. Can students tell firms they want

---

1. U.S. Food and Drug Administration, *New Era of Smarter Food Safety*, https://www.fda.gov/food/new-era-smarter-food-safety (last updated Apr. 29, 2021).
2. Emily M. Broad Leib & Baylen J. Linnekin, *Food Law & Policy: An Essential Part of Today's Legal Academy*, 13 J. Food L. & Pol'y 228, 230 (2017).
3. *Id.* at 260.

to go into "food law" as a defined field of practice? Even for practicing lawyers, wrapping their arms around the contours of this nebulous field can be difficult at first.

The goal of this practical guide to food law is to offer attorneys of all stripes an introduction to how different areas of law and legal practice intersect with food. Ranging from intellectual property to food policy and food regulation, this guide aims to whet the appetite for individuals looking to engage with the unique needs of clients in the food and beverage industry.

Food law, by its very nature, is both interdisciplinary and multidisciplinary.[4] Indeed, the vastness of the subject of "food law" leads some to question whether this area is merely a subfield of other areas of law, whether sounding in torts, consumer protection, administrative law, animal welfare, or some combination of any number of other legal domains.[5] However, the challenges facing food and beverage clients in today's modern food system require creative thinking, pragmatic approaches, and a focus on how food is regulated; that is to say, to best serve food and beverage clients, well-prepared lawyers should aim to meet these clients' particular needs.[6] Often these clients are seeking legal counsel who understand their business and are willing to dig into the details by, among other things, communicating risk mitigation strategies, consulting with experts in medical and nutrition science, and appreciating how their product is differentiated from competitors.

Food law is truly a unique field. It has the advantage of being eminently approachable for one simple reason: Everyone eats. As consumers ourselves, we each bring to the table our own perspectives on food. We each have made memories around the dinner table. Lawyers are central to assisting companies meeting their compliance obligations in getting food safely from farm to table and counseling businesses in helping consumers make informed choices about their food. In addition, attorneys assist individuals struggling with food insecurity to receive and maintain public benefits, especially in administrative appeals, to ensure that families obtain benefit levels to which they are entitled.

Put simply, food law is an approachable topic. In pieces, it may be easy to digest, but all at once, it appears hard to swallow. For readers new

---

4. Baylen J. Linnekin & Emily M. Broad Leib, *Food Law & Policy: The Fertile Field's Origins and First Decade*, 2014 WIS. L. REV. 557, 586–87 (2014).
5. MICHAEL T. ROBERTS, FOOD LAW IN THE UNITED STATES 10 (2016).
6. *See id.*

to the field of food law in its many forms, this practical guide aims to provide an overview of what lawyers actually do to assist food and beverage clients across various domains of legal practice. It is written by practicing lawyers *for* practicing lawyers, with a focus on information that is both practical and actionable.

Reader, this guide is intended to be your entrée into this engaging field of practice. Bon appétit!

Acknowledgments

Many people deserve thanks for their roles in this book's creation.

I thank the authors, who contributed their time and insight to the chapters contained herein. This project would not be possible without their contributions, borne from their experience. The finished product reflects your legal acumen and collective expertise. Thank you.

I also thank the American Bar Association (ABA), including my editor, Sarah (Sam) Forbes Orwig, and the good people of ABA Publishing. Thank you to Sam for keeping this project on track despite the pandemic, for your encouragement, and for all of your assistance throughout this process. Thank you also to the ABA Science & Technology Law Section and its Book Publishing Board who approved this proposal. It has been a privilege to be part of the ABA SciTech Section, and I thank you for your vote of confidence in this book idea. I encourage all attorneys reading this work to consider ABA membership and contributing to your area of interest via ABA sections and publications.

Thank you to my Perkins Coie colleagues, including David T. Biderman, Charles C. Sipos, Carrie Akinaka, and Larry Reichman. Thank you for your contributions to this work, and thank you for welcoming me to the firm after my lateral transition in 2019. David and Charles have built the country's most developed practice group in food litigation, and thank you for enabling me to grow as a lawyer and contribute to this growing field. Thank you also to Kathleen (Katie) O'Sullivan and James Williams for supporting my work with the ABA. Thanks also to Rebecca Grube and Justin Cole of Perkins Coie for all of your assistance during these many months! I also thank Susan Davies, Amanda Reiss, Andrew Mokey, Erin Krampetz, Cindy Squires, and others for their assistance in my career development. Thank you to Clicky Stone, Gavin Hilgemeier, and the Environmental Enforcement Section of the U.S. Department of Justice for cementing my decision to go to law school. Further, I wish to thank the Honorable Max O. Cogburn, Jr., David Davis, and Kathy Lancaster for a clerkship experience that was the best transition to legal practice that a new lawyer could dream of—thank you for the training and lessons that I will carry throughout my career.

Thank you to Michael T. Roberts and the University of California, Los Angeles (UCLA) Resnick Center for Food Law & Policy for graciously writing the foreword to this volume and for inviting me to serve on the Center's Advisory Board. Thank you also for the opportunity to teach a seminar on Food Litigation at UCLA Law; the students are a constant source of inspiration. Thank you to the staff of Ec10: Principles of Economics, including David Johnson, Anne Le Brun, and Paul Kelso, for the opportunity to teach while in law school.

Thank you to the champions of food law and policy, such as Professor Roberts, Peter Barton Hutt, and Emily Broad Leib. It is because of the tireless efforts of these champions that food law and policy has charted its own course and continues its path into a defined field of legal academia and practice. My time with Harvard's Food Law and Policy Clinic are some of my favorite memories in law school.

I thank the educators who I have had the good fortune to encounter throughout my educational journey. Special thanks to Dr. Michael V. Woodward (whose first assignment was "How to Read"), Dr. Pamela Childers, Skeeter Makepeace, Jason Jones, Bob Bires, Bryan Smith, Susie Brubaker-Cole, Steve Bartlett, Luther Killian, Rev. Ed Snodgrass, Sandra Carter, Alice Petty, John Pearson, Morris Graves, Richard Parker, Chris Robichaud, Chris McLaughlin, Judge David Barron, Valerie Jones Taylor, Juliet Aiken, William Dement, Barton J. Bernstein, and Adrienne Jamieson. Further thanks to Tammy Frisby and David M. Kennedy for your mentorship throughout my most significant pre-law school research project. I am greatly indebted to Martha Chavez and Susannah Barton Tobin for their mentorship and support over many years. Further, I wish to thank Albert Bandura, OC. Professor Bandura is not only a tremendous mind and a kind man, but he has influenced my life in more ways than I could possibly express. Professor Bandura, the zeal with which you approach your work is truly inspiring, as is your boundless optimism. May the efficacy force always be with you, Professor.

I would be remiss if I did not thank you, the reader. Thank you for your time and interest in the growing field of food law. On behalf of each of our contributors, welcome to this emerging area of practice.

Finally, but far from least, thank you to my family. Thank you to my father who throughout his long struggle has never given up hope. Thank you to my brother whose personal and professional success came after such persistence worthy of Professor Bandura's self-efficacy theory. Thank you to my mother, to whom this work is dedicated. Thank you for always being there and for teaching us that "food is love." Mom, thank you for *everything*.

Foreword

Michael T. Roberts
Executive Director, Resnick Center for Food Law & Policy
Professor from Practice
UCLA School of Law

Eighteen years ago, when I left my law practice to venture off to the University of Arkansas to join the law faculty and commence a life-long commitment to the building of food law and policy, a senior leader in the firm said to me "you are throwing away your career—no one talks about food or agriculture anymore; they are dead topics." A half-dozen years later, another senior leader of the firm who I ran into in an airport stated, "I thought you were making a mistake, but every time I pick up a newspaper, I read an article about food. How did you know food was going to be so important?" My answer was simply "I didn't know. Sometimes it is better to be lucky than smart."

I am lucky. I have enjoyed a multifaceted career in food law—from the University of Arkansas to a food-law practice in Washington, D.C., and Los Angeles to the University of California, Los Angeles (UCLA), where for the past seven years I have enjoyed seeing firsthand the growth of food law both in academia and practice. Food law has given me a pass to see the world, as academic and legal work in this space has literally taken me around the planet—from Asia to Europe to North America.

Food law has also introduced me to a cadre of creative, enterprising lawyers who have figured out how to make a living by posting the proverbial shingle of "food law," sometimes attached to another specialty like class action litigation or intellectual property law or international law—or, sometimes, just food law. I am pleased to see that the authors in this wonderful publication, *Food Law: A Practical Guide,* many of whom I know, are some of the leading lights in the practice of food law. I am just as pleased that there are a few authors who I don't know as well, which is further testament to the expansion and growth in this space.

When I taught my first food law and policy class 17 years ago, I had a distressing thought as I tried to fall asleep the night before class: what if some smart student asks me to define "food law"? What is its scope? What are its boundaries? I have confronted this same question when writing a treatise on food law in 2016 and in later publications and forums. It occurred to me at some point that food law is a little bit of this and a little bit of that. Put in more sophisticated language and to borrow language from the treatise:

> The vastness of "food law" subject matter raises a legitimate question as to whether food law is not a discipline in and of itself, but merely a subsection of other forms of law—administrative, environmental, consumer protection, international, tort, zoning, animal welfare, constitutional, and intellectual property. However, the case to consider food law as a discipline (albeit multidoctrinal) in and of itself is strong. Its value lies in focusing attention on how law governs food from the field to the table. The challenges posed by a modern food system unlike anything that the world has experienced have generated attention on the governance of food to warrant a legal field and discipline such as food law. Moreover, by recognizing how law governs food, improvements can be made and dynamics can be better understood.[1]

The collection of topics covered in this guide certainly underscore both the vastness of food law as well as the need to focus on this unique, multidoctrinal approach to a sector that is constantly in flux. I have often heard that food law is unruly; perhaps this is true and okay at the same time. The important point is that food law provides a forum or place where legal specialists can meet and in interconnected ways contribute to understanding and, in some cases, improving how food is governed.

I am thrilled that this publication offers this sort of forum for the private practitioner and offers a "practical" approach. As a scholar and some-time practitioner of food law, I fully appreciate the need for lawyers to share and exchange pointers on the practice. I am also impressed that the guide is a project of the American Bar Association's Science & Technology Law Section: in my opinion, the future of food law and policy is integrally connected to science and technology. Organizing the guide with this focus ensures that it will have a long shelf life.

On a personal note, I would like to acknowledge the skills and foresight of Tommy Tobin, the editor of this guide. Tommy and firm partner

---

1. MICHAEL T. ROBERTS, FOOD LAW IN THE UNITED STATES 10 (2016).

David Biderman of Perkins Coie LLP are teaching a seminar course at UCLA Law School this semester on food litigation and serve on the outside advisory board for the UCLA Resnick Center for Food Law & Policy. Tommy has an amazing amount of energy and capacity and I am grateful that much of this is directed towards the development of food law.

For the lawyer who is already part of this growing field of food law and for the lawyer who wishes to enter at some point, whether by complete immersion or as a complement to another area of practice, this guide should be an enduring, indispensable tool.

Chapter 1

Regulation of Food Formulation, Manufacturing, Labeling, and Advertising: A Primer, Farm to Fork

Ricardo Carvajal

Introduction

Food has become one of our most highly regulated commodities, and with good reason. No other type of product so directly and consistently touches on consumers' health and sense of well-being. Further, food is uniquely vulnerable to a range of abuses that can compromise its integrity, quality, and safety—a vulnerability that consumers cannot readily guard against, especially in a marketplace that is increasingly globalized.

Throughout the nation's early history, regulation of food was a function left to the states. The emergence of national markets, together with widespread adulteration of food and fraudulent marketing practices, gave rise to a movement that spurred enactment of the 1906 Pure Food and Drug Act (Pure Food Act) and the Federal Meat Inspection Act (FMIA). The Pure Food Act contained basic adulteration and misbranding provisions and granted authority for case-by-case enforcement through seizure and prosecution. The FMIA provided for pre-slaughter inspection, post-mortem examination, application of an inspection mark, and destruction of condemned meat.

Although the 1906 Pure Food Act marked a significant advancement, its limits soon became apparent. For example, in *United States v.*

Lexington Mill & Elevator Co.,[1] the U.S. Supreme Court made clear that the government bore the burden of showing the relationship between a substance found in food and any alleged harm—a difficult showing, especially given the limited scientific methods available at the time. In addition, economic adulteration of food continued to be a significant problem.

Congress responded through passage of the 1938 Federal Food, Drug, and Cosmetic Act (FDCA). That law authorized factory inspections, provided for the establishment of food standards to help ensure the integrity of basic staples such as wheat flour, and added injunction as a remedy. Administered by the U.S. Food and Drug Administration (FDA, or the agency), the FDCA remains the principal law governing federal regulation of food safety and labeling. Over the subsequent decades, the FDCA has been amended to address additional challenges as they emerged. The following are a few of those major amendments:

- **1954.** Concerns over the use of chemicals in food led to the Miller Pesticide Amendment, which provided for setting limits on pesticide residues in raw agricultural commodities.
- **1958.** The 1958 Food Additives Amendment provided a mechanism for pre-market control over substances added to food, which placed the burden of demonstrating the safety of any proposed use on the proponent of that use. It also provided a mechanism for the use of safe levels of poisonous or deleterious substances.
- **1990.** The rise of nutrition science and a better understanding of the links between certain nutrients and certain diseases led Congress to pass the Nutrition Labeling and Education Act (NLEA). That law required nutrition labeling and authorized pre-market review of claims that characterize the level of a nutrient in food as well as claims that describe the relationship between a nutrient and a disease.
- **1994.** Controversy and uncertainty over the regulation of products marketed as dietary supplements led to passage of the Dietary Supplement Health and Education Act, which made clear that dietary supplements are subject to regulation as a type of food and established certain requirements specific to that category of foods.
- **2002.** Passed in the wake of the terrorist attacks on September 11, 2001, the Bioterrorism Act required that food facilities register with FDA, required that firms keep records of their immediate

---

1. 232 U.S. 399 (1914).

suppliers and subsequent recipients (referred to as "one up, one down") to enhance traceability, and required prior notice of imported food shipments.
- **2004**. The Food Allergen Labeling and Consumer Protection Act established specific requirements for labeling of major food allergens to better enable food-allergic consumers to avoid foods that could pose a hazard to their health.
- **2011**. In response to continuing outbreaks of foodborne illness and the increasing globalization of the food supply, Congress passed the Food Safety Modernization Act (FSMA)—an ambitious overhaul of the nation's food safety system that shifted industry's and the government's focus from response to prevention. Implementation of that law has been a herculean effort that continues to unfold.

As an important aside, in the course of the evolution described above, FDA shifted from heavy reliance on case-by-case enforcement to heavy reliance on the issuance of regulations and guidance documents—a development that tracks the evolution of administrative law. That phenomenon helps explain the relative dearth of more recent case law, and the explosion of regulations that began in the 1970s and has continued unabated.

Federal Agencies That Regulate Food

FDA

FDA is the principal agency that oversees regulation of food. The agency consists of various components with discrete functions. For example, the Center for Food Safety and Applied Nutrition provides scientific and policy support regarding regulation of human foods (including dietary supplements) and color additives. The Center for Veterinary Medicine does the same with respect to animal foods and also with respect to animal drugs, which must be used in accord with requirements that ensure the safety of any food products derived from the animal. These product centers work closely with the Office of Regulatory Affairs, which takes the lead on activities in the field such as factory inspections, review of products offered for import, and recalls. All of these agency components report to the Office of the Commissioner, which provides strategic leadership for the agency, and draws on legal advice from the Office of the Chief Counsel.

Pursuant to the FDCA, FDA has jurisdiction over "food," which is broadly defined to mean (1) articles used for food or drink for man or other animals, (2) chewing gum, and (3) articles used for components of any such article.[2] As we will see, this broad definition can encompass practically anything used or destined for use as or in food, including live cattle, most seafood, eggs in the shell, microbes used in food production, and chemicals such as carbon dioxide used for carbonated beverages. That said, some federal courts have put their own gloss on "food" as something that is consumed "primarily for taste, aroma, or nutritive value."[3] For most regulatory purposes, dietary supplements are considered to be "food."

Many states have their own food laws, which generally parallel the FDCA. The states have regulatory agencies that serve as counterparts to FDA, and with which FDA coordinates many of its activities; in fact, most FDA inspections are conducted by state regulators under contract with FDA. Regulation of the retail food sector is primarily left up to counties and localities, but FDA aims for consistency and coordination through publication of the Food Code. Although the Food Code itself is not binding, it has been adopted into law in many jurisdictions.

U.S. Department of Agriculture

The U.S. Department of Agriculture (USDA), through its Food Safety and Inspection Service (FSIS), administers the FMIA, Poultry Products Inspection Act, and Egg Products Inspection Act. Pursuant to those laws, FSIS exercises jurisdiction over products derived from certain species of food animals and over processed egg products. FSIS assumes jurisdiction over animals once they enter the slaughterhouse. Meat and poultry processing establishments operate under continuous inspection by FSIS—a considerably more stringent level of oversight than FDA's system of periodic inspections. FSIS also exercises much tighter control on importation of meat and poultry products, which must come from foreign establishments designated as eligible for exporting to the United States. Finally, FSIS has authority to review product labels prior to marketing, whereas FDA generally reviews labels in the context of post-market inspections.

Alcohol and Tobacco Tax and Trade Bureau

The Alcohol and Tobacco Tax and Trade Bureau (TTB), an agency within the U.S. Department of the Treasury, has primary jurisdiction over most alcoholic beverages under the Federal Alcohol Administration Act (FAA

---

2. 21 U.S.C. § 321(f).
3. Nutrilab v. Schweiker, 713 F.2d 335, 338 (7th Cir. 1983).

Act).[4] Although TTB regulates labeling of such beverages, TTB relies on FDA for help with safety evaluations in the context of product contamination and ingredient authorization. Some alcoholic beverages fall outside the scope of the FAA Act, and FDA has primary jurisdiction over those beverages. For example, FDA has primary jurisdiction over beers that fall outside the FAA's definition of "malt beverage," which requires the use of both malted barley and hops. Thus, such beers must comply with all applicable requirements under the FDCA and FDA's implementing regulations.

U.S. Environmental Protection Agency

The U.S. Environmental Protection Agency (EPA) registers pesticides under the authority of the Federal Insecticide, Fungicide, and Rodenticide Act. As part of the registration process, EPA establishes tolerances for pesticide residues in food. Under the authority of the FDCA, FDA enforces the tolerances established by EPA. Also, FDA has jurisdiction over antimicrobials used in or on food during the course of food processing and manufacture.

Federal Trade Commission

The Federal Trade Commission (FTC) regulates food advertising under the authority of the Federal Trade Commission Act (FTCA). That law prohibits unfair or deceptive acts or practices, as well as advertising that is false or misleading in a material respect. In many instances, there will be an overlap between FTC and FDA because FDA has jurisdiction over "labeling," which is broadly defined to mean "all labels and other written, printed, or graphic matter (1) upon any article or any of its containers or wrappers, or (2) accompanying such article."[5] As a practical matter, FDA's primary interest often lies in ensuring that a food product is properly labeled and not marketed with claims that render the food an unapproved drug, whereas FTC's primary interest lies in ensuring that advertisers have a reasonable basis for claims used in advertising.

Food Formulation

A fundamental regulatory requirement for food formulation is that any *use* of a substance in food must be permitted by law. It follows that one

---

4. TTB also regulates taxation of alcohol, a function that lies beyond the scope of this summary.
5. FDCA § 201(m).

use of a substance in food may be permissible, whereas a different use of the substance may not. This is made clear by the legal definition of a "food additive" as

> any substance *the intended use of which* results or may reasonably be expected to result, directly or indirectly, in its becoming a component or otherwise affecting the characteristics of any food (including any substance intended for use in producing, manufacturing, packing, processing, preparing, treating, packaging, transporting, or holding food; and including any source of radiation intended for any such use).[6]

Food Additive Approval

If the use of a substance meets the definition of a food additive (which presumes that no exception applies), then that use must be the subject of a petition submitted to and approved by FDA. FDA's approval results in the issuance of a regulation listing the conditions of safe use of the additive.[7] Approval is conditioned on a showing that the use of the additive is "safe," meaning that "there is a reasonable certainty in the minds of competent scientists that the substance is not harmful under the conditions of its intended use."[8] Further, FDA cannot approve a petition if the additive is found to induce cancer when ingested by humans or animals, or if it is found to induce cancer in humans or animals based on appropriate safety testing—a proviso referred to as the Delaney Clause, in remembrance of the legislator who pressed for its inclusion in the 1958 Food Additives Amendment. Review and approval of a petition can be expected to take at least two years, and the approval is not exclusive to the petitioner. There is a separate, faster pre-market notification and review process for food additives that also qualify as food contact substances, meaning substances used in packaging materials. That process results in an authorization that is exclusive to the notifier.

"Generally Recognized as Safe" Exception

There are several exceptions to the definition of "food additive." The most important is referred to as the "generally recognized as safe" (GRAS) exception, which applies when a substance is generally recognized among appropriately qualified experts as having been adequately

---

6. *Id.* § 201(s) (emphasis added).
7. 21 C.F.R. pts. 172–180 (2021).
8. *Id.* § 170.3(i).

shown to be safe under the conditions of its intended use.[9] There are two bases for application of the GRAS exception: (1) common use in food prior to January 1, 1958, and (2) scientific procedures.

A claim to GRAS status based on common use in food requires a demonstration of "a substantial history of consumption of a substance for food use by a significant number of consumers."[10] The history of use must be documented and can be based on experience in other countries. In practice, adequately documenting a substantial history of consumption prior to 1958 can be exceedingly difficult, such that most conclusions of GRAS status in the modern era are based on scientific procedures.

A claim to GRAS status based on scientific procedures must be based on scientific information that demonstrates the safety of the proposed use of the substance. The term "scientific procedures" is flexibly defined to include "the application of scientific data (including, as appropriate, data from human, animal, analytical, or other scientific studies), information, and methods, whether published or unpublished, as well as the application of scientific principles, appropriate to establish the safety of a substance under the conditions of its intended use."[11]

Regardless of the basis for GRAS status, the evidence relied on must be in the public domain—in other words, generally available to appropriately qualified experts. Otherwise, the "general recognition" element of the GRAS standard will not be satisfied. Further, there must exist a basis to conclude that the evidence is generally accepted among appropriately qualified experts—in other words, that there exists a consensus about safety. Consensus does not mean unanimity, and the existence of consensus can be confirmed through a variety of means, including peer-reviewed scientific journals, secondary scientific literature, the opinion or recommendation of an authoritative scientific body, or the opinion of a panel of appropriately qualified experts.

The FDCA permits a company to reach its own conclusion of GRAS status of the use of a substance, and there is no requirement for FDA review and approval of that conclusion. However, if FDA later disagrees with the company's conclusion—either with respect to safety or general recognition of safety—then FDA could deem any food containing the substance to be adulterated as a matter of law because the food contains a

---

9. There are several other exceptions to the definition of "food additive." These include exceptions for pesticide chemicals, pesticide chemical residues, new animal drugs, color additives, and dietary ingredients used in dietary supplements.
10. *Id.* § 170.3(f).
11. *Id.* § 170.3(h).

food additive that is "unsafe" (i.e., unapproved). Introducing adulterated food into interstate commerce or adulterating a food in interstate commerce is a prohibited act. Further, adulterated food is subject to seizure and, in the case of imports, to refusal of admission.

Given this risk—and usually also based on commercial considerations—some companies will avail themselves of the opportunity to voluntarily submit a notice of their GRAS conclusion to FDA for the agency's review.[12] The time frame for FDA's review of the notice once the notice has been accepted is six months. If FDA decides that it has no questions regarding the company's conclusion, FDA will issue a letter stating as much. If the agency raises an issue during its review that the company is not able to resolve in a timely manner, then the company usually will choose to withdraw its notice, and FDA will issue a letter summarizing the issue and acknowledging the withdrawal. FDA posts both the company's notice and the agency's response on the agency's website.

If a substance is intended for use in a meat or poultry product, then it must clear the additional hurdle of an FSIS review of suitability for the proposed use. In this context, suitability refers to the effectiveness of the substance, and whether its use could result in an adulterated or misleading product. FSIS's review is governed by a memorandum of understanding between USDA and FDA.[13]

The hurdle of securing food additive approval or establishing GRAS status is not the only one that a company has to clear at the formulation stage. Additional potential hurdles include (1) FDA's fortification policy, which sets out principles to guide food fortification, and discourages fortification of certain categories of foods (e.g., snack foods such as candies and carbonated beverages);[14] (2) applicability of one of the many standards of identity established by FDA, which can prescribe product composition, manufacture, and labeling;[15] and (3) applicability of section 301(ll) of the FDCA, which prohibits the introduction into interstate commerce of any food to which has been added an approved drug or licensed biologic, or a drug or biologic for which substantial clinical investigations

---

12. *See* 21 C.F.R. pt. 170, subpt. E (human food) and pt. 570, subpt. E (animal food) (2021).

13. Memorandum of Understanding between the Food Safety and Inspection Service, U.S. Department of Agriculture, and the Food and Drug Administration, U.S. Department of Health and Human Services, MOU 225-00-2000 (last amended Jan. 15, 2015), https://www.fda.gov/about-fda/domestic-mous/mou-225-00-2000-amendment-1.

14. 21 C.F.R. § 104.20 (2021).

15. *Id.* pts. 130–169.

have been instituted and made public, unless the drug or biologic was marketed in food before approval or licensing, and before substantial clinical investigations were instituted.

Dietary Supplements

In the case of dietary supplements, the considerations described above apply in determining the regulatory status of ingredients other than dietary ingredients—in other words, ingredients that play the same supporting role in a dietary supplement as they play in a conventional food (e.g., an anticaking agent or flavor). As for dietary ingredients, they must first be demonstrated to qualify as such. The term "dietary ingredient" refers to any one of several categories listed in the statutory definition of "dietary supplement," namely (1) a vitamin; (2) a mineral; (3) an herb or other botanical; (4) an amino acid; (5) a dietary substance for use by man to supplement the diet by increasing the total dietary intake; or (6) a concentrate, metabolite, constituent, extract, or combination of any of the preceding ingredients.[16] A dietary supplement can contain one or more such dietary ingredients.

If a dietary ingredient was not marketed in the United States before October 15, 1994, then it is a "new dietary ingredient" (NDI).[17] Unless an NDI is present in the food supply as an article used for food in a form in which the food has not been chemically altered, then the manufacturer must submit a notification to FDA that documents the basis for the manufacturer's conclusion of safety under the conditions of intended use. The notification must be submitted at least 75 days prior to marketing. If FDA raises concerns that the manufacturer cannot satisfactorily resolve, FDA will issue a letter summarizing those concerns. Eventually, both the notification and FDA's response are made public. The failure to submit a required notification deems a dietary supplement adulterated.

As with conventional foods, there are other hurdles to consider at the formulation stage. These include whether the product is intended for ingestion—a fundamental requirement—and the potential applicability of section 201(ff)(3)(B) of the FDCA. That provision parallels section 301(ll) (discussed above), and excludes from the definition of "dietary supplement" an approved drug, certified antibiotic, or licensed biologic, or any such article for which substantial clinical investigations have been instituted and made public, unless there exists evidence of prior marketing as a dietary supplement or food.

---

16. FDCA § 201(ff)(1).
17. *Id.* § 413(d).

In summary, assessing and confirming the regulatory status of a substance intended for use in food can be a complex endeavor that can entail some uncertainty and that can require collaboration between the supplier of the substance and the manufacturer of a product formulated to contain that substance. The cost and time required to carry out this vital task should be factored into product development.

Food Manufacture: Ensuring Safety

Depending on the product, the manufacture of food can be a complex endeavor drawing on ingredients acquired through far-flung supply chains and calling for multiple processing steps. It follows that there are multiple opportunities for the introduction of contaminants and other problems that can compromise the safety and quality of food.

Historically, efforts to regulate food safety were primarily a reactive endeavor, and were undertaken pursuant to bedrock provisions of the FDCA that FDA still relies on today. These provisions define basic ways that a food can be deemed adulterated and are worth discussing in some detail.

Section 402(a)(1) of the FDCA is the provision that FDA relies on when a food is demonstrably contaminated and potentially injurious. Under that provision, a food is adulterated if it bears or contains an added poisonous or deleterious substance that "may render [the food] injurious to health." Under the "may render" standard, FDA asks whether there is a reasonable possibility that the food has been rendered injurious to the health of consumers, including those most vulnerable.[18] Potential adulterants generally fall into one of three categories: microbiological (e.g., *Listeria monocytogenes*, *Salmonella* spp., *E. coli* (esp. 0157:H7), *Vibrio vulnificus*, *C. botulinum*); chemical (e.g., mercury and other heavy metals, melamine); and physical (choking hazards, glass, metal).

If a poisonous or deleterious substance is inherent (as opposed to added), then a food will not be deemed adulterated if the quantity of the substance does not ordinarily render the food injurious to health. Under the "ordinarily injurious" standard, FDA asks whether the food is injurious to the health of consumers under ordinary conditions of use. For example, Japanese star anise contains toxins that can induce severe inflammation and epileptic seizures, and it is therefore adulterated.[19] In comparison, foods to which some consumers are allergic (e.g., peanuts,

---

18. United States v. Lexington Mill & Elevator Co., 232 U.S. 399 (1914).
19. By contrast, Chinese star anise is safe for consumption.

eggs) are not considered to be ordinarily injurious to health and are not considered adulterated under this provision.

If a substance is both inherent but also present in part due to human agency, then the entire amount is considered "added," and is evaluated under the more protective "may render injurious" standard. For example, mercury is present in the environment both as a result of volcanic activity and human activity. Therefore, all mercury in seafood is considered "added."[20] Similarly, a fungal toxin called aflatoxin can occur naturally in corn. However, naturally occurring levels can be increased through post-harvest mishandling, such that the entire amount would be considered "added."

If FDA finds significant sanitation problems at a food storage or processing establishment, FDA can act to prevent food from that establishment from entering commerce until the problems are resolved—even if the food is not demonstrably contaminated.[21] Under section 402(a)(4) of the FDCA, a food is adulterated if it has been prepared, packed, or held under insanitary conditions whereby it may have become contaminated with filth or been rendered injurious to health. FDA need not show that the food is actually contaminated; evidence of insanitary conditions giving rise to the potential for contamination is sufficient to deem all of the food processed or held under those conditions to be adulterated as a matter of law.

Finally, under section 402(a)(3) of the FDCA, FDA can deem a food adulterated if the agency finds it to consist in whole or in part of *any* filthy, putrid, or decomposed substance, *or* if it is otherwise unfit for food. Because this provision is exceedingly broad and unforgiving, FDA has established action levels for certain types of defects in certain foods.[22] These serve as points of reference against which to evaluate whether defects in a given food are consistent with industry norms, based on the application of good manufacturing practice.

Not all of FDA's food safety efforts have been reactive. In 1969, FDA relied on its authority under section 402(a)(3) and (4) (discussed above) to promulgate a general Current Good Manufacturing Practice (CGMP) regulation that applied broadly across the food industry and

---

20. United States v. Anderson Seafoods, Inc., 622 F.2d 157 (5th Cir. 1980).

21. FDCA § 402(a)(4).

22. *See* FDA, Food Defect Levels Handbook (2018), *available at* https://www.fda.gov/food/ingredients-additives-gras-packaging-guidance-documents-regulatory-information/food-defect-levels-handbook.

that addressed various common elements of sanitation.[23] The regulation includes requirements and recommendations relating to personnel (disease control, sanitation); plant and grounds (design, maintenance, drainage); sanitary operations (storage of toxic substances, pest control, cleaning food contact surfaces); sanitary facilities and controls (water supply, sewage and trash disposal, hand-washing and toilet facilities); equipment and utensils (design and maintenance); production processes and controls (raw materials, temperature control, protection against contamination); and warehousing and distribution (protection against contamination in storage and transportation). In subsequent years, FDA established additional, more specific CGMP regulations for certain types of foods, tapping other statutory authorities as needed.[24]

In the 1970s, FDA relied on the authority granted under section 404 of the FDCA to establish specific regulatory requirements for the processing of thermally processed low-acid foods packaged in hermetically sealed containers, as well as the processing of acidified foods. Improper processing of these foods can give rise to formation of botulinum toxin by certain species of bacteria in the genus *Clostridium*. The requirements governing processing of low-acid foods are set forth in 21 C.F.R. part 113, and those governing processing of acidified foods are set forth in part 114. Both of these regulations are referenced in part 108, which sets out the process for FDA's issuance of an order that requires a processor to obtain an emergency permit before any further commercial distribution of a low-acid or acidified food. Although FDA frequently alleges violations of these regulations, FDA has rarely invoked its authority to require a processor to obtain an emergency permit.

In the 1990s, FDA took a more marked turn toward a preventive approach by working to establish Hazard Analysis and Critical Control Points (HACCP) requirements for the seafood and juice sectors. Under an HACCP approach, a manufacturer must conduct a hazard analysis to identify hazards that are reasonably likely to occur, and then develop and implement a plan to control those hazards. Under the plan, the manufacturer must identify critical control points for each hazard, identify critical limits to be met at each control point, set up monitoring at each control point, take corrective action when there is deviation from a critical limit, verify the effectiveness of the plan, and keep extensive records

---

23. This "umbrella" CGMP regulation was at 21 C.F.R. part 110 but is now at part 117, subpart B.

24. For example, for infant formula in 21 C.F.R. part 106, dietary supplements in part 111, and bottled water in part 129.

demonstrating compliance. In effect, the manufacturer is expected to anticipate potential food safety problems and prevent them before they occur, and to make appropriate adjustments when there are failures. The HACCP regulations for seafood at 21 C.F.R. part 123 and for juice at part 120 are still in effect.

The turn toward a preventive approach took a dramatic leap in 2011 when FSMA was signed into law.[25] Passage of that legislation was propelled by recognition that a preventive approach to food safety needed to be extended across the food industry. The law spurred FDA to issue seven regulations that the agency refers to as "foundational":

- Preventive Controls Rule for Human Food
- Preventive Controls Rule for Animal Food
- Produce Safety Rule
- Foreign Supplier Verification Programs (FSVP) Rule
- Accredited Third-Party Certification Rule
- Sanitary Transportation Rule
- Intentional Adulteration Rule

This massive rulemaking effort required years of sustained engagement by FDA, its state counterparts, and industry, and will require still more time for full implementation. A detailed review of the rules is well beyond the scope of a survey chapter such as this, so the following paragraphs are intended only to provide a high-level glimpse of what the first four foundational rules entail.[26]

The Preventive Controls Rule for Human Food at 21 C.F.R. part 117 updated the general CGMP regulation and expanded on the approach in FDA's HACCP regulations. The rule applies to domestic facilities, and also foreign facilities that manufacture, process, pack, or hold food for consumption in the United States. Each such facility must have a written food safety plan prepared by, or under the oversight of, an appropriately qualified individual. The plan must include a written hazard analysis from which other obligations may flow, depending on the outcome of the hazard analysis. This requirement ensures that relevant hazards are significantly minimized or prevented, and that food is not adulterated under section 402 of the FDCA or misbranded under section 403(w) (which governs labeling of major food allergens).

---

25. Pub. L. No. 111-353, 124 Stat. 3885.
26. All of the regulations and associated guidance documents and other resources are readily available on FDA's website.

The hazard analysis requires a manufacturer to identify known or reasonably foreseeable hazards. A "hazard" is "any biological, chemical (including radiological), or physical agent that has the potential to cause illness or injury."[27] A "known or reasonably foreseeable hazard" is "a biological, chemical (including radiological), or physical hazard that is known to be, or has the potential to be, associated with the facility or the food."[28] A hazard can be naturally occurring, unintentionally introduced, or intentionally introduced for economic gain.

Once the manufacturer has identified known or reasonably foreseeable hazards, then the manufacturer must evaluate those hazards to determine whether they are hazards that require a preventive control. That determination necessarily calls for the exercise of judgment based on knowledge and experience. More specifically, the regulation defines a hazard requiring a preventive control as one

> for which a person knowledgeable about the safe manufacturing, processing, packing, or holding of food would, based on the outcome of a hazard analysis . . . establish one or more preventive controls to significantly minimize or prevent the hazard in a food and components to manage those controls . . . as appropriate to the food, the facility, and the nature of the preventive control and its role in the facility's food safety system.[29]

If the manufacturer determines that a hazard requires a preventive control, then the manufacturer must choose and apply the appropriate control. The regulation flexibly defines "preventive controls" as

> those risk-based, reasonably appropriate procedures, practices, and processes that a person knowledgeable about the safe manufacturing, processing, packing, or holding of food would employ to significantly minimize or prevent the hazards identified under the hazard analysis that are consistent with the current scientific understanding of safe food manufacturing, processing, packing, or holding at the time of the analysis.[30]

---

27. 21 C.F.R. § 117.3 (2021).
28. *Id.*
29. *Id.*
30. *Id.* As with the HACCP approach, preventive controls include controls at critical control points, or CCPs (meaning "a point . . . at which control can be applied and is essential to prevent or eliminate a food safety hazard"). However, preventive controls also include "[c]ontrols, other than those at CCPs, that are also appropriate for food safety." *Id.* § 117.135(a)(2)(ii). Thus, the preventive controls approach can be viewed as an expansion of the HACCP approach.

There are a number of different types of controls that might be appropriately included in a food safety plan, including process controls, food allergen controls, sanitation controls, and supply chain controls. Once these have been implemented, they must be monitored to ensure that they are consistently performed. If there is a lapse, then the facility must take corrective action and prevent affected food from entering commerce. Further, the facility must conduct verification activities to ensure that the preventive controls are validated and that the system is working as intended, and must keep comprehensive records and make them available to FDA.

There are a number of at least partial exemptions in the preventive controls rule. For example, there is an exemption for dietary supplement facilities that comply with CGMP requirements and serious adverse event reporting requirements that are specific to dietary supplements. There is an additional exemption for facilities solely engaged in storage of raw agricultural commodities (other than fruits and vegetables) intended for further distribution or processing or storage of packaged foods that are not exposed to the environment. There is also an exemption for farms, which turns on the rather complex definition of "farm" in 21 C.F.R. § 1.227. Finally, certain facilities—including those that qualify as a very small business—are subject to modified requirements that are intended to impose lower regulatory burdens.

The second foundational rule, Preventive Controls Rule for Animal Food at 21 C.F.R. part 507, is very similar to the Preventive Controls Rule for Human Food. The principal differences arise from the fact that hazards for animals may differ from those for humans. For example, undeclared food allergens are not generally considered a hazard for animals. Also, evaluation of potential hazards for animals must take the target species into account because different species of animals can be vulnerable to different hazards. Finally, foods for companion animals such as dogs and cats can present hazards to humans because they are brought into the home and are subject to more direct handling than food intended for consumption by livestock.

The third foundational rule is the Produce Safety Rule at 21 C.F.R. part 112. This rule is intended to mitigate microbiological hazards associated with domestic and imported fruits and vegetables that are consumed raw. In the years preceding the passage of FSMA, there had been repeated outbreaks of foodborne illness associated with those foods. The toll associated with those outbreaks was such that FDA economic analysis of the Produce Safety Rule estimated that its implementation would result in a decrease of 362,059 illnesses per year, valued at $976 million.

Because farms are somewhat of a new constituency for FDA, the agency faced numerous challenges in developing the rule. FDA invested significant effort to identify conditions and practices that were potential contributing factors to foodborne illness. Ultimately, the agency opted to include requirements specific to the following major areas:

- Agricultural water (e.g., establishment of microbial criteria for irrigation and post-harvest wash)
- Use of biological soil amendments of animal origin
- Worker health and hygiene
- Equipment, tools, buildings, and sanitation (design, cleaning, and maintenance)
- Assessment of potential for contamination by domesticated and wild animals
- Growing, harvesting, packing, and holding activities
- Production of sprouts (a commodity associated with numerous outbreaks)

Produce safety is one area in which FDA has relied especially heavily on state agencies to help implement the requirements in their jurisdictions.

With the fourth foundational rule, the FSVP Rule at 21 C.F.R. part 1, subpart L, FDA extended its regulatory reach over importers. The overarching goal of this rule is to ensure that imported food is as safe as domestic food. To that end, importers must verify that imported food complies with any applicable preventive controls and is not adulterated under section 402 of the FDCA, or misbranded under section 403(w). Note that the importer for purposes of the FSVP Rule may or may not be the same person as the importer of record for customs purposes. Whereas an importer of record may be the owner or purchaser of the goods, or a designated licensed customs broker, the FSVP regulation defines "importer" as the U.S. owner or consignee at the time of entry, or the U.S. agent of a foreign owner or consignee at the time of entry.[31] Being designated as the FSVP importer entails significant added responsibilities, as the regulation requires the FSVP importer to evaluate and approve suppliers, and to conduct appropriate verification activities that can include on-site audits, review of a supplier's food safety records, and periodic testing and sampling of shipments. It is thus crucial that there be clear agreement on which of the entities involved in the importation of a food will serve as the FSVP importer.

---

31. *Id.* § 1.500.

No discussion of food manufacture and food safety would be complete without at least a brief mention of what happens when there is a lapse that results in the distribution of food that fails to meet applicable requirements and, worse yet, does so in a way that gives rise to a possibility of injury. Depending on the probability and severity of the hazard associated with the food, a manufacturer may be required to submit a report to FDA and may also be expected or required to conduct a recall. The manufacturer should promptly investigate the matter to determine whether the food qualifies as a "reportable food," meaning "an article of food (other than [dietary supplements or] infant formula) for which there is a reasonable probability that the use of, or exposure to, such article of food will cause serious adverse health consequences or death to humans or animals."[32] If so, then the manufacturer must submit a report to FDA through the agency's Reportable Food Registry within 24 hours of the manufacturer's determination.

Regardless of whether a food qualifies as a reportable food, the manufacturer may be expected to conduct a voluntary recall of the food. The depth and breadth of that recall will depend on a variety of factors, including the probability and severity of the hazard associated with the food. FDA has a regulation setting forth the agency's recall policy and procedures. Pursuant to that regulation, FDA will classify a recall as Class I ("a situation in which there is a reasonable probability that the use of, or exposure to, a violative product will cause serious adverse health consequences or death"), Class II ("a situation in which use of, or exposure to, a violative product may cause temporary or medically reversible adverse health consequences or where the probability of serious adverse health consequences is remote"), or Class III ("a situation in which use of, or exposure to, a violative product is not likely to cause adverse health consequences").[33] As a very rough rule of thumb, a Class I recall typically will extend to the consumer level, a Class II recall to the retail level, and a Class III recall to the wholesale level.

If the manufacturer refuses to conduct a voluntary recall, and if the situation meets the Class I criteria, then FDA has authority to mandate the conduct of the recall.[34] Because the vast majority of companies choose to voluntarily conduct a recall under the agency's supervision, FDA has rarely invoked its mandatory recall authority.

---

32. FDCA § 417(a)(2).
33. 21 C.F.R. § 7.3(m)(1)–(3) (2021).
34. FDCA § 423.

In appropriate cases, FDA can also pursue judicial remedies, including seizure and injunction. In egregious cases, FDA has undertaken criminal investigations and, if warranted, has sought criminal prosecution of companies and individuals. Lawyers advising food companies should be aware that the FDCA is a strict liability statute that provides for civil and criminal penalties.[35]

Labeling and Advertising

When used in reference to an FDA-regulated product, the terms "label" and "labeling" have specific legal definitions. In relevant part, a "label" is "a display of written, printed, or graphic matter *upon the immediate container of any article*."[36] This definition is consistent with common usage. However, the term "labeling" has a broader definition that has implications that are not readily apparent. "Labeling" is "all labels and other written, printed, or graphic matter (1) upon any article or any of its containers or wrappers, *or* (2) *accompanying such article*."[37] In the modern era, FDA interprets the latter phrase to encompass websites through which a product is sold or referenced on the label of a product. Thus, the concept of "labeling" can overlap with the concept of "advertising," which is generally understood to encompass promotional materials in virtually any format. This means that the practitioner should consider the legal implications of labeling and advertising from multiple regulatory perspectives, including those of FDA, FTC, and their state counterparts.

From the regulators' perspective, the role of labeling and advertising is to clearly communicate any required information to the consumer and to ensure that any information that is voluntarily provided is truthful and not misleading. From the marketer's perspective, labeling and advertising are important drivers of sales. This sets up a tension that frequently results in regulatory challenges to product labeling and advertising by federal and state regulators, which can trigger follow-on civil litigation that can be expensive to defend. Notwithstanding that expense, compliance with regulatory requirements can be somewhat of an afterthought for some companies, especially newer entrants to the market.

---

35. *Id.* §§ 301 and 303(a).
36. *Id.* § 201(k) (emphasis added).
37. *Id.* § 201(m) (emphasis added).

Mandatory Labeling Requirements

FDA's labeling requirements are extensive and detailed and can be complex to apply. Certain requirements are mandatory and relatively straightforward. For example, a product's principal display panel (PDP)—defined as the panel most likely to be viewed under ordinary conditions of purchase—must bear a statement of identity and the net quantity of contents. These elements are critical to a consumer's purchasing decision, as they tell the consumer what the product is and how much of the product is in the container. The statement of identity must be an appropriately descriptive term, and might already be specified by law, or can be determined by reference to principles articulated in FDA regulations. The statement of identity is distinct from whatever brand or trade name the marketer chooses to use, if any.

Other required information must appear on the PDP or the information panel, which is usually immediately to the right of the PDP. That includes the nutrition facts (required by the NLEA and very tightly regulated with respect to content and format); a declaration of ingredients by common or usual name (with an emphasis on technical accuracy, as opposed to marketing cachet); disclosure of major food allergens; and the name and place of business of the manufacturer, packer, or distributor. In some instances, a violation of these requirements will be minor, and of little consequence. In other instances—such as failure to appropriately disclose the presence of a major food allergen—the consequences can be severe, both in terms of costs associated with corrective action and potential injury to the consumer.

Voluntary Labeling Requirements

Other information that a marketer chooses to voluntarily provide can be subject to strict regulatory requirements. For example, a statement that characterizes the level of a nutrient in food is considered a nutrient content claim (NCC). An NCC can only be used in accord with criteria established by FDA through its regulations; otherwise, the food is deemed misbranded.[38] Some NCCs signal that a product contains dietarily significant amounts of a desired nutrient (e.g., "good source of vitamin C"), whereas others signal that a product contains lower levels of a less desirable nutrient (e.g., "low salt"). Some words—such as "healthy"—can carry regulatory implications because FDA views them as implied NCCs, depending on the context in which they are used.

---

38. *Id.* § 403(r)(1)(A); 21 C.F.R. § 101.13 and subpt. D (2021).

As an additional example of a voluntary claim that triggers strict regulatory requirements, a statement that characterizes the relationship between a food substance and a disease or health-related condition is considered a health claim.[39] Failure to use a health claim in accord with criteria established by FDA can not only deem a food misbranded but could also render it an unapproved new drug. Because of their potential health significance, health claims generally must be supported by significant scientific agreement (a very high standard), and must be authorized prior to use. If there is a lack of significant scientific agreement, it is nonetheless possible to petition FDA for the use of a qualified health claim, meaning a claim that is qualified in a way that accurately communicates the underlying level of scientific support to the consumer. FDA's regulation of health claims has been the subject of extensive litigation grounded in First Amendment jurisprudence—a fascinating tangent that lies beyond the scope of this chapter but would be a worthwhile detour for the intellectually curious.[40]

Other types of voluntary claims are not burdened with such extensive and detailed regulatory requirements, but nonetheless can carry significant legal implications. For example, a claim regarding a product's or ingredient's effect on a structure or function of the body is permissible. Such claims are referred to as structure/function claims. When used in reference to a dietary supplement, a structure/function claim must be notified to FDA within 30 days of its first use. When used in reference to a conventional food (meaning a food other than a dietary supplement), a structure/function claim does not have to be notified to FDA; however, the structure/function effect must derive from the food's nutritive value (e.g., a value in sustaining human existence by such processes as promoting growth, replacing loss of essential nutrients, or providing energy). In any case, because a structure/function claim does not require pre-market authorization and is not the subject of specific qualifying criteria,[41] a

---

39. *See* FDCA § 403(r)(1)(B); 21 C.F.R. § 101.14 and subpt. E (2021).

40. While health claims can be made with appropriate authorization, other claims that reference a disease usually will render a food an unapproved new drug. The pandemic has seen a flurry of FDA warning letters objecting to products marketed for the prevention or treatment of COVID-19. *See* FDA, *Fraudulent Coronavirus Disease 2019 (COVID-19) Products*, https://www.fda.gov/consumers/health-fraud-scams/fraudulent-coronavirus-disease-2019-covid-19-products (last updated Apr. 27, 2021).

41. But note that FDA's regulation at 21 C.F.R. § 101.93 sets out factors that FDA considers in determining whether a structure/function claim implies treatment or prevention of disease, in which case the claim would be considered a disease claim that renders the product an unapproved new drug.

marketer might be tempted to use it without fully understanding that the claim must still be truthful and not misleading. To that end, FTC requires that claims used in advertising be adequately substantiated *prior to their first use*.[42]

The requirement of adequate substantiation is significantly thornier than might first appear. Whether a given claim is adequately substantiated—that is, whether there is a reasonable basis for the claim—depends on the nature of the claim. For example, if the claim can be considered a claim that has a bearing on consumers' health or safety, then the claim must be supported by a high level of substantiation in the form of "competent and reliable scientific evidence." This standard has been defined as "tests, analyses, research, studies, or other evidence based on the expertise of professionals in the relevant area, that have been conducted and evaluated in an objective manner by persons qualified to do so, using procedures generally accepted in the profession to yield accurate and reliable results."[43] FTC frequently finds health- and safety-related claims—including structure/function claims—to lack this requisite level of support.

Some voluntary claims might be the subject of standards administered by a government agency or a third-party organization. For example, the term "organic" is subject to regulation through the National Organic Program administered by the Agricultural Marketing Service of USDA. Failure to use that term in accord with applicable requirements can leave a marketer vulnerable to financial penalties and civil litigation. As another example, the term "non-GMO," to distinguish a product that is not derived from a genetically engineered organism, is the subject of a standard administered by the Non-GMO Project, a nongovernmental organization. Conformance with that standard enables a marketer to use the "Non-GMO Project Verified" seal on qualifying products. Although the claim "non-GMO" can be used independent of verification by a third party, conformance to a third-party standard can help boost credibility of the claim and potentially ward off challenges grounded in an alleged lack of substantiation.

Finally, claims used in labeling and advertising have increasingly become the focus of consumer litigation, especially in states viewed

---

42. FTCA section 5 prohibits "unfair or deceptive acts or practices." Section 12 prohibits dissemination of any "false advertisement," which section 15 defines to include an advertisement that "is misleading in a material respect." *See* FTC Enforcement Policy Statement on Food Advertising (May 13, 1994).

43. *See, e.g., In re* Schering Corp., 118 F.T.C. 1030 (1994).

as friendly to plaintiffs. Typically, those cases are framed as violations of state laws governing consumer fraud and other deceptive or unlawful practices. Frequently, the cases arise in the wake of an enforcement action by FDA or FTC because those actions are made public. Although there is no private right of action under the FDCA, plaintiffs have proven adept at creatively leveraging alleged violations of the FDCA into alleged violations of state law. Numerous defenses may be available in such cases, including defenses grounded in explicit and implied preemption and in the doctrine of primary jurisdiction, as well as the defenses more typically employed in class action litigation. However, given the cost of mounting such a defense, investing adequate resources in ensuring regulatory compliance is a worthwhile investment.

Concluding Thoughts

Entities and individuals that choose to engage in the business of producing and distributing food must do so with the awareness that they are entering a highly regulated market that requires constant vigilance to maintain product safety, quality, and integrity. The legal landscape for food has evolved significantly over the past century, and the past few decades in particular have seen a significant increase in regulatory complexity. The practitioner willing to invest the time and effort to master that complexity will be rewarded with opportunities to serve as a trusted and invaluable resource to a wide range of businesses that operate at all points in the continuum from farm to fork.

Chapter 2

Food Litigation: An Emerging Field

Tommy Tobin

Food litigation has quickly become a distinct practice area in commercial litigation, with the plaintiffs' bar increasingly targeting the industry, law firms establishing practice teams focusing on these matters, and even law schools offering courses on the subject.[1] Since 2015, the nation's courts have seen at least 750 new food litigation filings.[2] Even though overall civil litigation filings were down considerably due to the pandemic in 2020, the number of food litigation cases actually rose by more than 20 percent.[3] This growing field sees no signs of slowing down in the years ahead.

This chapter discusses several considerations specific to food litigation practice. Additionally, an overview of some of the fundamentals of civil practice, particularly the types of claims that often affect food businesses, class action basics, and common defenses to claims, is provided.

Food and Civil Litigation

Just as food law entails a set of sometimes disparate sources of law that "combine to regulate the market (and politics) of food: who eats what and

---

1. *See* UCLA Law, *Law 693—Food Litigation: Consumer Protection, Regulation, and Class Actions*, https://curriculum.law.ucla.edu/Guide/Course/5448 (last visited May 4, 2021).
2. Perkins Coie, Food & Consumer Packaged Goods Litigation: 2020 Year in Review 4 (2021).
3. Charles Sipos et al., *Courts Are Right to Scrutinize Food Labeling Suits*, Law360, Aug. 4, 2020, https://www.law360.com/articles/1297119/courts-are-right-to-scrutinize-food-labeling-suits.

why,"[4] so too does food litigation entail a variety of legal approaches. Food companies are regulated at the federal, state, and local levels. At the federal level, federal agencies, such as the U.S. Food and Drug Administration (FDA), U.S. Department of Agriculture, and Federal Trade Commission (FTC), oversee varying aspects of the food system from manufacturing standards to food safety protocols to labeling claims. At the state level, states have enacted state consumer protection acts and often their own labeling requirements, both of which may provide a private right of action for individual consumers. Regulatory actions, whether through warning letters, recalls, or enforcement actions, can prompt the plaintiffs' bar to bring putative class action cases alleging consumer injury.[5]

The complex network of federal, state, and even local laws applicable to food businesses can result in compliance risks, especially for market innovators who attempt to differentiate their product from competitors. Civil disputes involving food businesses can take many forms, from food safety product liability matters to false advertising claims brought by rival companies.

Consumer class actions in the food industry are among the most recent and important trends in civil litigation generally. Especially since 2015, plaintiffs' attorneys are increasingly targeting food businesses on a class-wide basis.

Rise of Food Litigation

Recent years have seen a remarkable rise in food litigation cases, as demonstrated by the table below.[6] In the years since 2008, food litigation filings have exceeded 1,400 new matters.

Year	Food Litigation Filings	Year-on-Year Increase (%)
2008	19	—
2009	26	7 cases (37%)
2010	45	19 cases (73%)
2011	53	8 cases (18%)
2012	95	42 cases (79%)
2013	94	-1 case (-1%)
2014	81	-13 cases (-14%)

---

4. Michael T. Roberts, Food Law in the United States 4 (2016).
5. *Id.* at 303.
6. Perkins Coie, *supra* note 2, at 4.

Year	Food Litigation Filings	Year-on-Year Increase (%)
2015	158	77 cases (95%)
2016	145	-13 cases (-8%)
2017	145	0 cases (0%)
2018	164	19 cases (13%)
2019	179	15 cases (9%)
2020	220	41 cases (23%)

Several notable trends emerge from the data.

- **Massive growth after 2014.** New food litigation filings nearly doubled between 2014 and 2015, jumping to more than 150 putative class actions a year. While many reasons are possible for the drastic increase in food litigation suits between 2014 and 2015, many point to the U.S. Supreme Court's decision in *AT&T Mobility LLC v. Concepcion,* upholding contractual limitations on consumers' ability to pursue class actions in arbitration.[7] Others point to the active role of FDA and FTC among other regulators to monitor the marketplace and an increased prominence of consumer group investigations and media campaigns.[8]
- **Food litigation filings continue to grow.** Food litigation cases rose more than 20 percent between 2019 and 2020, even with the pandemic. But even as filings against the food industry continue to increase, the state and federal judiciaries continue to scrutinize the substance of such cases.[9]
- **California and New York are common venues.** While food litigation cases are filed in courts across the country, California and New York courts see the lion's share of new matters. The Northern District of California is even sometimes called the nation's "food court," given the number of cases filed there. In 2019, the two states saw roughly similar amounts of new cases, with California courts seeing 69 filings and 75 filed in New York's courts. In 2020, by contrast, New York courts saw a whopping 107 filings, with California courts seeing 58. The increase was largely because New York has been the predominant jurisdiction

---

7. Erica A. Burgos, *Selling the Footlong Short: How Consumers Inch toward Satisfaction in Costly Food Class Action Litigation*, 13 SEVENTH CIRCUIT REV. 259, 266 (2017).

8. ROBERTS, *supra* note 4, at 303.

9. Sipos et al., *supra* note 3.

for a spate of recent cases challenging products flavored with vanilla—presenting an idiosyncratic theory adopted by one plaintiffs' lawyer whose "vanilla" cases were consistently tossed out by federal courts in 2020.[10]

- **Flavors de jour.** Trends in food litigation filings fluctuate over time. For example, slack-fill filings, alleging that there were inappropriate levels of nonfunctional space in the product's packaging, comprised 23 percent of all new food litigation cases in 2016. But in 2020, these were only 4 percent of the total of new cases. Similarly, cases alleging issues with a food product's "natural" or "all natural" representation totaled more than 20 percent of cases filed in 2015–2018 but represented 6 percent of filings in 2019. A significant recent trend is the considerable increase of "vanilla" cases.[11] These cases make up more than a quarter of all filings in 2019 and 2020.

Yet, the data masks certain realities. For example:

- **Pre-litigation discussions.** The compiled data on *filed* food litigation cases does not reflect the many cases that are threatened, but never actually filed. Many state consumer protection statutes, such as California's Consumers Legal Remedies Act,[12] require a pre-suit notice and demand prior to the filing of an action brought under the statute. Ostensibly, these demand letters offer companies an opportunity to cure the alleged issue. More often, these letters provide an opportunity for each side's counsel to address the core of a consumer's complaint *prior* to any court filing and to potentially negotiate a mutually agreed resolution.
- **Negative publicity from public filings.** Food companies, and their associated brands, are some of the most recognizable and beloved companies and brands in the country. Food litigation filings often make a splash, as plaintiffs contend that merely lodging a public lawsuit against a well-known food company is a newsworthy event, regardless of whether the claim has any merit. By default, public court filings are available to the public, and sometimes might take a company by surprise.

---

10. PERKINS COIE, *supra* note 2, at 4.
11. Corinne Ramey, *The Case for Plain Vanilla Gets Its Day in Court*, WALL ST. J., Feb. 7, 2021, https://www.wsj.com/articles/the-case-for-plain-vanilla-gets-its-day-in-court-11612724626.
12. *See* CAL. CIV. CODE § 1750 *et seq.*

- **Case length and complexity.** Especially as these cases proceed, food litigation matters can be lengthy, expensive, and highly complex, involving federal standards, state consumer protection law, fact-specific inquiries, and class action procedure.

Types of Cases

As litigation involving food companies can take many forms, businesses and their counsel should be familiar with common claims affecting the industry. Given its prominence as a vehicle to get putative class plaintiffs' complaints into court, food businesses should pay particular attention to state consumer protection laws.

Consumer Protection Statutes

Consumer protection laws exist in all 50 states. Generally, they protect the public against misleading and unfair business practices, such as false advertising. Many of these laws are enforceable by the state attorneys general as well as private parties.

Prior to exercising private rights of action, individuals must often first comply with procedural requirements, such as sending a pre-suit notice to the defendant company.[13]

While the exact requirements of the applicable consumer protection law depend on the statute's text and its interpretation in the relevant state courts, these laws generally require that (1) a consumer is (2) injured or deceived and that damage was (3) caused by (4) defendant's deceptive or misleading act or trade practice. Important distinctions exist as to the elements of consumer protection laws, and local laws must be consulted. One of the most critical distinctions is whether the laws require that a consumer relied upon the allegedly deceptive or misleading act.[14]

Some states, such as New York and California, have multiple consumer protection laws in the same state that are often litigated together, but they nonetheless may contain varying requirements and present different available remedies, especially with regard to injunctive relief or

---

13. *See, e.g.,* ALA. CODE § 8-19-10(e) (requiring advance notice); MASS. GEN. LAWS ch. 93A, § 9(3) (generally requiring pre-suit notice); Consumers Legal Remedies Act (CLRA), CAL. CIV. CODE § 1782 (same).

14. *Compare* ARK. CODE § 4-88-113(f) (requiring reliance as precondition to private right of action) *with* Odom v. Fairbanks Mem'l Hosp., 999 P.2d 123, 132 (Alaska 2000) (noting that under Alaska consumer protection law, "Actual injury as a result of the deception is not required. . . . All that is required is a showing that the acts and practices were capable of being interpreted in a misleading way").

statutory damages.[15] Unless prohibited by state law,[16] state consumer protection statutes are often used as vehicles for putative class action plaintiffs to get their cases into court. Federal laws, notably the Federal Food, Drug, and Cosmetic Act, do not provide a private right of action for consumers to address labeling claims via the court system; however, many plaintiffs nonetheless plead violations of state consumer protection law to get the same or similar concerns before a court.

Products Liability and Negligence

Food manufacturers should take all reasonable care to ensure that products they make and sell are safe for consumers to consume. Sometimes, even though a company has exercised all reasonable care, a product might nonetheless cause injury to consumers. FDA identifies that foods made under unsanitary conditions, for example, may be "adulterated" under federal law as the food could be injurious to health.[17]

In product liability actions, consumers might allege that there was a defect in the product's design, manufacturing, or warnings that contributed to it becoming unreasonably dangerous for consumption.[18] In order to prevail, consumers would also need to demonstrate that the defect caused the injury claimed. State laws on strict products liability have significant distinctions that should be taken into account.

In negligence actions, plaintiffs must establish the familiar five elements of duty, breach, cause-in-fact, proximate cause, and injury. Food companies should be aware that they may have a duty to exercise reasonable care in supplying goods for consumers' use, especially as that product might be consumed.[19] In terms of proximate cause, companies

---

15. *See* N.Y. GEN. BUS. LAW §§ 349 and 350; CLRA, CAL. CIV. CODE § 1750 *et seq.*; False Advertising Law, CAL. BUS. & PROF. CODE § 17500 *et seq.*; Unfair Competition Law, CAL. BUS. & PROF. CODE § 17200 *et seq.*

16. *See* ARK. CODE § 4-88-113(f)(1)(B) ("A private class action under this section is prohibited unless the claim is being asserted for a violation of Arkansas Constitution, amendment 89."); LA. REV. STAT. § 51:1409(A) (consumer "may bring an action individually but not in a representative capacity to recover actual damages"); MONT. CODE § 30-14-133(1) (consumer "may bring an individual but not a class action" under state consumer protection law).

17. *See* 21 C.F.R. § 110.5.

18. *See* RESTATEMENT (SECOND) OF TORTS § 402A.

19. *See generally* Edward C. v. City of Albuquerque, 241 P.3d 1086 (N.M. 2010) ("Two categories of legal duty are recognized: (1) an affirmative duty to conform one's actions to a specific standard of care in relation to a specific individual or group of individuals created by a specific statutory or common-law standard; and (2) a

should consider the intended and advertised uses of the product and how consumers are likely to interact with a food product.

California's Proposition 65

California's Safe Drinking Water and Toxic Enforcement Act of 1986, known more commonly as Proposition 65, requires businesses that sell consumer products—including food—notify Californians about certain chemicals that are in those products that are known to the state to cause cancer or reproductive health issues.[20] Given that California is a prime market for food companies, manufacturers and sellers should take care to label products appropriately when marketing products in that state.

Violations of Proposition 65 labeling requirements can pose substantial litigation risks, especially as Proposition 65 allows for a private right of action. Failure to comply with warning requirements can result in statutory violations of $2,500 per violation per day.

Like the rise of food litigation filings more generally, Proposition 65 pre-suit notices targeting the food and beverage industry have risen considerably between 2015–2020, as shown in the table below.

Year	Proposition 65 Pre-Suit Notices Filed
2015	56
2016	160
2017	385
2018	382
2019	413
2020	1,116 (as of November 2020)

Each year, there are hundreds of Proposition 65 filings and related settlements with many repeat plaintiff organizations and plaintiffs' counsel. The coming years are likely to see even more notices filed by these repeat players.

Other Laws and Legal Requirements

Other laws and legal requirements of particular concern for food companies from a litigation perspective include the following.

---

defensive duty that is the general negligence standard, requiring the individual to use reasonable care in his activities and dealings in relation to society as a whole.").

20. *See* Tommy Tobin et al., *An Introduction to California's Proposition 65: What Is Proposition 65?*, FOOD LITIG. NEWS, July 27, 2020, https://www.foodlitigationnews.com/2020/07/an-introduction-to-californias-proposition-65-what-is-proposition-65/.

Class Action Requirements

Given the risk of exposure to consumer class actions, food companies, and their counsel, should be aware of the requirements necessary to certify a class. Federal courts recognize that the class certification determination might sound "the death knell of the litigation."[21]

Federal Rule 23(a) provides for four key criteria for class certification. Each requirement must be satisfied. The elements of class certification required by Federal Rule 23(a) are:

- **Numerosity**. The number of individuals must make individual joinder impractical.
- **Commonality**. The members of the class must present common questions of law or fact.
- **Typicality**. The claims or defenses of the class representatives must be typical of those of class members.
- **Adequacy**. The class representatives must fairly and adequately represent the interests of the class.

Slack-Fill Cases

In slack-fill cases, plaintiffs allege that nonfunctional space is actionable under state consumer protection law. These cases often hinge on the interpretation of federal standards on slack-fill, namely 21 C.F.R. § 100.100. Under that regulation, "a food shall be deemed to be misbranded if its container is so made, formed, or filled as to be misleading," including for reasons of *nonfunctional* slack-fill.[22] Defendants' consistent wins at the federal level appear to have prompted fewer plaintiffs to bring these cases in recent years.

Other Theories

Food litigation filings also attempt many other theories, including:

- **False labeling**. This category encompasses a broad range of theories: under- or over-reporting of the amount of a nutrient or ingredient in a product (e.g., level of sugar or protein); allegations that the food was not made in the manner suggested by the label (e.g., foods labeled "smoked" deriving their smoke flavor from chemicals versus smoke); referring to ingredients as "real" when they are allegedly processed (e.g., "real cocoa"); and misstating the number of servings in a container (e.g., instant coffee

---

21. Chamberlan v. Ford Motor Co., 402 F.3d 952, 957 (9th Cir. 2005).
22. 21 C.F.R. § 100.100(a).

yielding fewer servings than the label indicated, when used as instructed). The breadth and variation of these cases demonstrate how plaintiffs' counsel continue to scrutinize all label claims for new angles on potential civil liability.[23]

- **"All natural."** These cases allege, among other things, that the presence of some non-zero amount of a synthetic molecule, such as glyphosate, renders the product's "natural" or "all natural" claim false or misleading. Even with a "natural" claim, courts have found that reasonable consumers "would not be so absolutist as to require that 'natural' means there is no glyphosate, even an accidental and innocuous amount" in their products.[24]
- **Animal welfare and environmental practices.** As consumer packaged goods companies continue to focus on claims that communicate a product's or brand's environmental benefits and animal welfare practices, the plaintiffs' bar appears to be focusing on such claims as an emerging area of litigation. Several recently filed cases have alleged that the treatment of animals in the manufacturing process or claims of environmental stewardship rendered product labeling and advertising false or misleading.
- **Place of origin.** These cases allege that the words and imagery on product labels misled consumers into believing that the product was made in a certain location or contained ingredients sourced from a particular location. A subset of these cases claim that the product name itself is misleading to consumers.
- **Vanilla.** Between 2019 and 2020, more than 100 cases were filed, primarily by the same individual attorney, generally alleging the same core claim: a "reasonable consumer" expects that a product labeled as having been flavored with vanilla cannot derive *any* of its flavor from other sources beyond pure vanilla or vanilla extract. Recent cases have indicated an increasing impatience by the federal courts toward this onslaught of vanilla litigation, as courts in New York dismissed complaints challenging vanilla-flavoring claims on at least four occasions over a 12-month period.[25]
- **Competitor litigation.** Whether through the Lanham Act or the National Advertising Division (NAD) of the Better Business

---

23. PERKINS COIE, *supra* note 2, at 5.
24. Parks v. Ainsworth Pet Nutrition, LLC, 377 F. Supp. 3d 241 (S.D.N.Y. 2019).
25. PERKINS COIE, *supra* note 2, at 9.

Bureaus, competitors might challenge false advertising.[26] One of the few food litigation matters to reach the Supreme Court, *POM Wonderful LLC v. Coca-Cola Co.*, was a Lanham Act case.[27] While the Lanham Act might see competitors litigating in federal courts, the NAD may be a more cost-effective and efficient process through a self-regulatory system.

Common Defenses

Food companies defending lawsuits have numerous defenses available to them, which vary by particular claim and jurisdiction. Below is a review of common defenses, particularly in the class action context.

Reasonable Consumer

Pursuant to Federal Rule 12(b)(6), a complaint that fails to state a claim for which relief can be granted should be dismissed. At the motion to dismiss stage, a federal court generally accepts well-pled allegations as true, but claims must cross the line from conceivable to plausible in order to escape dismissal.[28]

In consumer class action suits, members of the public must be "likely to be deceived" by the alleged misrepresentation.[29] Courts evaluate whether the public is "likely to be deceived" under the reasonable consumer standard. Where the pleading does not plausibly allege that a reasonable consumer would be deceived, the action should not survive dismissal.[30]

The reasonable consumer defense is an important tool for food companies looking to defend consumer class action cases.[31] In consumer class action matters generally, and food litigation matters specifically, the reasonable consumer defense is often dispositive. Strained and implausible

---

26. *See* Nitika Arora et al., *When to Choose between a Lawsuit or Filing a Challenge with the NAD*, ADWEEK, Aug. 30, 2019, https://www.adweek.com/agencies/when-to-choose-between-a-lawsuit-or-filing-a-challenge-with-the-nad/.

27. 573 U.S. 102 (2014).

28. Bell Atl. Corp. v. Twombly, 550 U.S. 544, 570 (2007).

29. *See* Becerra v. Dr Pepper/Seven Up, Inc., 945 F.3d 1225, 1228 (9th Cir. 2019); *accord* Ebner v. Fresh, Inc., 838 F.3d 958, 965 (9th Cir. 2016).

30. *See, e.g.*, Painter v. Blue Diamond Growers, 757 F. App'x 517, 519 (9th Cir. 2018); Cruz v. Anheuser-Busch Cos., LLC, 682 F. App'x 583, 583 (9th Cir. 2017); Carrea v. Dreyer's Grand Ice Cream, Inc., 475 F. App'x 113, 115 (9th Cir. 2012).

31. PERKINS COIE, FOOD LITIGATION 2019 YEAR IN REVIEW 9 (2020), https://www.perkinscoie.com/images/content/2/2/229474/2019-Food-Litigation-YIR-v4.pdf.

readings of disputed labeling terms are unlikely to survive a reasonable consumer review.[32] For example, many courts understand that a reasonable consumer exercises "common sense" and a plaintiff's *unreasonable* interpretation of a key label phrase is not actionable. As the Ninth Circuit recently put it, "a reasonable consumer does not check her common sense at the door of the store."[33]

Pleading with Specificity

Federal Rule 9(b) requires special pleading standards in cases alleging fraud or mistake. The pleadings must state these claims with particularity. Federal courts have interpreted Rule 9(b) to require that a complaint "(1) specify the statements that the plaintiff contends were fraudulent, (2) identify the speaker, (3) state where and when the statements were made, and (4) explain why the statements were fraudulent."[34] In short, Rule 9(b) requires the who, what, when, and how of a fraud allegation.

Food companies facing allegations that consumers were misled by a purportedly fraudulent claim must have notice of what the claim was, where it was located, how the claim was communicated to the plaintiff, and why the claim is false. This information will assist the company in responding to and defending against the lawsuit's allegations.

Preemption

Generally, preemption defenses identify that a higher legal authority, such as the federal government, so binds the lower authority as to be superior when these authorities conflict. Given the many layers of regulation governing food products, preemption is an important consideration.

Two species of preemption exist: express and implied. Express preemption occurs when a higher authority directly opposes the lower authority. Implied preemption occurs when the higher legal authority, generally the federal government, creates a conflict between complying with the two authorities or occupies the field in that area of law.

Litigants may be able to point to compliance with federal or state standards to preempt consumer protection or tort claims. A leading case on federal preemption in the food and beverage context is *Turek v.*

---

32. *See id.*
33. Weiss v. Trader Joe's, No. 19-55841, 2021 WL 816075, at *1 (9th Cir. Mar. 3, 2021); *accord* Daniel v. Mondelez Int'l, Inc., 287 F. Supp. 3d 177, 193 (E.D.N.Y. 2018) ("A reasonable consumer does not lack common sense.").
34. Rombach v. Chang, 355 F.3d 164, 170 (2d Cir. 2004).

General Mills, Inc.,[35] which found that the plaintiff's claims regarding fiber content on a food product were preempted under federal law.

Primary Jurisdiction

Primary jurisdiction is a prudential doctrine leading the courts to defer to agencies for an initial determination of a contested issue. The primary jurisdiction doctrine applies where claims implicate a federal agency's expertise with a regulated product, and the court wishes to take advantage of the agency's expertise in order to promote uniformity and the integrity of the regulatory scheme surrounding the regulated product.[36]

Four factors guide the court's review of whether to apply the primary jurisdiction doctrine: (1) the need to resolve an issue that (2) has been placed by Congress within the jurisdiction of an administrative body having regulatory authority (3) pursuant to a statute that subjects an industry or activity to a comprehensive regulatory scheme that (4) requires expertise or uniformity in administration.[37] The court's decision to apply primary jurisdiction and defer its decision making is discretionary.

Most notably, primary jurisdiction has been invoked over FDA's decision-making process involving the term "natural." As FDA has indicated that regulation of "natural" remains an ongoing agency priority, courts have continued to stay "natural" cases under the primary jurisdiction doctrine in deference to FDA's deliberative process.[38]

Elsewhere, courts have stayed several cases involving cannabidiol (CBD) based on the agency's statements since November 2020 that it was exploring "potential pathways for various types of CBD products to be lawfully marketed" and it expected to provide an update on its "progress regarding the agency's approach to these products in the coming weeks."[39]

---

35. 62 F.3d 423 (7th Cir. 2011).
36. Greenfield v. Yucatan Foods, L.P., 18 F. Supp. 3d 1371, 1375 (S.D. Fla. 2014).
37. *In re* Horizon Organic Milk Plus DHA Omega-3 Mktg. & Sales Practice Litig., 955 F. Supp. 2d 1311, 1348 (S.D. Fla. 2013).
38. Perkins Coie, *supra* note 31, at 8.
39. *Id.*; *see also* Tommy Tobin, *CBD in Food and Beverage: 3 Developments to Watch in 2020*, Forbes, Dec. 31, 2019, https://www.forbes.com/sites/tommytobin/2020/12/31/cbd-in-food--beverage-three-developments-to-watch-in-2020 ("After the agency's November announcement and warning letters, manufacturers of CBD products, particularly those in food and beverage categories, should watch how the FDA clarifies and enforces its position on CBD-containing products over the new year."); David Biderman et al., *Awaiting FDA Rulemaking, Courts Are Pressing Pause on CBD Class Actions*, Cannabis Bus. Executive, Aug. 3, 2020, https://www.cannabisbusinessexecutive.com/2020/08/awaiting-fda-rulemaking

Conclusion

The table is set for food litigation's continued growth in the coming years. As this case law continues to develop, food businesses should work proactively to identify potential litigation risks, and a substantial part of that process is understanding the potential claims and defenses companies are likely to encounter.

-courts-are-pressing-pause-on-cbd-class-actions/ ("CBD companies facing consumer class action suits regarding product labeling have a reasonable chance of success in staying the case on primary jurisdiction grounds while the FDA continues to contemplate its regulatory action. But this doctrine relies on litigants effectively linking the anticipated regulatory action with the issues involved in the case.").

Chapter 3

Reasoning with the Reasonable Consumer Standard in Food Litigation

Charles C. Sipos, Carrie Akinaka, and Tommy Tobin

Introduction

Picture this: Crunch Berries. The plaintiff claims that the bright colored cereal, shaped to resemble berries, paired with the term "Crunch Berries," misled her into thinking that the cereal actually contained berries.[1] She alleges that the product's packaging conveyed that this Cap'n Crunch product contained real, nutritious fruit and had she known that the product contained *no fruit* she would not have purchased the product over the prior four years. Should a court rely on the *subjective* view of this individual plaintiff in determining whether such a claim is actionable? Surely not. Enter the *objective* reasonable consumer standard, which requires plaintiffs to show a probability that a significant portion of consumers acting reasonably would be misled. Put differently, the reasonable consumer standard enables a court to call this Crunch Berries case what it is—"nonsense"—and to dismiss such a subjective, *unreasonable* claim at the motion to dismiss stage.[2]

---

1. *See* Sugawara v. PepsiCo, Inc., No. 208CV01335-MCEJFM, 2009 WL 1439115 (E.D. Cal. May 21, 2009).
2. Werbel *ex rel.* v. PepsiCo, Inc., No. C 09-04456 SBA, 2010 WL 2673860, at *3 (N.D. Cal. July 2, 2010) ("It is obvious from the product packaging that no

From torts to patents, and from privacy to employment, courts have fashioned legal standards based on a "reasonable person." The reasonable person standard reminds us that "people must conform to an objective standard of behavio[r]."[3] In tort law, for example, it is axiomatic that negligence is based on the objective reasonable person standard. Black's Law Dictionary defines a "reasonable person" as "a person who exercises the degree of attention, knowledge, intelligence, and judgment that society requires of its members for the protection of their own and of others' interests."[4]

False advertising law, too, has its own reasonable person: the reasonable *consumer*. With the rise of food litigation in recent years, the reasonable consumer standard has emerged as an important defense to claims that strain credulity. These cases generally do not focus on issues of product safety or integrity, but instead often make allegations regarding deception over labeling claims where no reasonable consumer could possibly be misled.[5] This chapter discusses some of the key debates bearing on the reasonable consumer standard.

Doctrinal Origins

While the reasonable person standard dates back hundreds of years,[6] the reasonable consumer standard has a somewhat more recent heritage. The Federal Trade Commission (FTC) was organized to regulate competition about a century ago. Among other things, the agency is authorized to monitor the marketplace regarding unfair and deceptive trade practices. Just what constitutes "unfair" or "deceptive" acts is undefined under the act creating the FTC, leaving it to the courts to determine.[7] Over time, states adopted "mini-FTC" acts and incorporated them into

---

reasonable consumer would believe that Cap'n Crunch derives any nutritional value from berries.").

3. R. v. Smith (Morgan), 1 A.C. 146, 172 (2000) (Lord Hoffman, C.J.).

4. *Reasonable Person*, BLACK'S LAW DICTIONARY (11th ed. 2019).

5. Charles Sipos et al., *Courts Are Right to Scrutinize Food Labeling Suits*, LAW360, Aug. 4, 2020, https://www.law360.com/articles/1297119/courts-are-right-to-scrutinize-food-labeling-suits.

6. *See Reasonable Person*, BLACK'S LAW DICTIONARY (11th ed. 2019).

7. This is in part by design. The U.S. Supreme Court noted in *Federal Trade Commission v. R.F. Keppel & Bro.*, 291 U.S. 304, 314 (1934), that it was: "It is unnecessary to attempt a comprehensive definition of the unfair methods which are banned, even if it were possible to do so . . . . New or different practices must be considered as they arise in the light of the circumstances in which they are employed." *See also* Pan Am. World Airways v. United States, 371 U.S. 296, 307–08 (1963) (finding that

state consumer protection law, and courts developed a body of case law (often varying between states) defining actionable "unfair" or "deceptive" conduct.

Meanwhile, the FTC itself wrestled with how to determine "unfairness" and "deception." In 1980, the FTC issued its Policy Statement on Unfairness, finding that the agency would evaluate two factors in assessing whether a business practice was unfair: (1) whether the practice injures consumers and (2) whether it violates established public policy.[8] Testifying in 1982, the FTC's chairman noted:[9]

> There are specific problems with the Commission's definition of deception. First, the definition is not clear, despite its 44 year history. The courts tend to give the Commission very wide latitude, and the Commission's own case law is not clear and consistent. *As a result, businesses do not know what they can and cannot do. Consumers do not know what protections they do and do not have.* The Commission really does not know what cases to bring and what not to bring, and the courts do not know which Commission decisions to affirm and which to reverse.

Subsequently, the FTC issued a formal Policy Statement on Deception, which incorporated the reasonable consumer standard, based upon the fact that consumers generally are capable of protecting themselves from unscrupulous trade practices.[10] The FTC set three criteria for assessing claims of deception: (1) there must be a representation, omission, or practice that is likely to mislead the consumer, (2) which is assessed from the perspective of a consumer acting reasonably under the circumstances, and (3) which is also "material," meaning that the act, representation, or practice is likely to affect the consumer's conduct regarding a product or service.[11] According to the FTC, a deceptive practice, omission, or representation "must be likely to mislead reasonable consumers under

---

the standard for what constitutes an unfair method of competition or unlawful practice must be determined on a case-by-case basis).

8. Public Statement, FTC, FTC Policy Statement on Unfairness (Dec. 17, 1980), https://www.ftc.gov/public-statements/1980/12/ftc-policy-statement-unfairness.

9. *FTC's Authority over Deceptive Advertising: Hearing before the Subcomm. for Consumers of the Senate Comm. on Commerce, Science, and Transp.*, 97th Cong. 3 (1982).

10. Jack E. Karns, *The Federal Trade Commission's Evolving Deception Policy*, 22 U. RICH. L. REV. 399, 411 (1988).

11. Public Statement, FTC, FTC Policy Statement on Deception (Oct. 14, 1983) [hereinafter FTC Policy Statement on Deception], https://www.ftc.gov/system/files/documents/public_statements/410531/831014deceptionstmt.pdf.

the circumstances" presented in a case when assessing "the totality" of that practice.[12]

The FTC, in its way, presaged the reasonable consumer standard's use as a tool courts might use to turn back silly claims involving food—claims driven either by ignorance or by creative lawyering by plaintiffs' counsel. Harkening back to agency commentary from the 1960s, the FTC explained:

> An advertiser cannot be charged with liability with respect to every conceivable misconception, however outlandish, to which his representations might be subject among the foolish or feeble minded. . . . Perhaps a few misguided souls believe, for example, that all "Danish pastry" is made in Denmark. Is it therefore an actionable deception to advertise "Danish pastry" when it is made in this country? Of course not.[13]

Rise of Food Litigation

Over time, courts have increasingly seen putative class action cases targeting the food industry. The rise in such cases has been nothing short of staggering, as demonstrated by the table below.[14] In the years since 2008, food litigation filings have exceeded 1,400 new matters.

Year	Food Litigation Filings	Year-on-Year Increase (%)
2008	19	—
2009	26	7 cases (37%)
2010	45	19 cases (73%)
2011	53	8 cases (18%)
2012	95	42 cases (79%)
2013	94	-1 case (-1%)
2014	81	-13 cases (-14%)
2015	158	77 cases (95%)
2016	145	-13 cases (-8%)
2017	145	0 cases (0%)
2018	164	19 cases (13%)
2019	179	15 cases (9%)
2020	220	41 cases (23%)

---

12. *Id.*
13. Heinz v. Kirchner, 63 F.T.C. 1282, 1290 (1963).
14. Perkins Coie, Food & Consumer Packaged Goods Litigation: 2020 Year in Review 4 (2021), https://www.perkinscoie.com/images/content/2/4/241153/2021-Food-CPG-Litigation-YIR-Report-v4.pdf.

As shown above, new cases nearly doubled between 2014 and 2015, jumping to more than 150 putative class actions a year. While many reasons are possible for the drastic increase in food litigation suits between 2014 and 2015, many point to the U.S. Supreme Court's decision in *AT&T Mobility LLC v. Concepcion*, upholding contractual limitations on consumers' ability to pursue class actions in arbitration.[15] Other factors likely include a substantial $21 million class settlement in 2010 between the FTC and state attorneys general involving Dannon's Activia yogurt and its claims of certain probiotic-related health benefits.[16] The Dannon settlement may have signaled to enterprising class attorneys that food litigation could pay out in a big way—well beyond the broken tooth, $5,000 a case, regime.

In the intervening years, as the body of case law has built up around food litigation class claims, the pace of filings has only accelerated. Even with the pandemic, 2020 was a record year for new food litigation filings.[17]

In consumer class action matters generally, and food litigation matters specifically, the reasonable consumer defense is often dispositive. For example, in the Crunch Berries case above, the case hinged on the plaintiff's subjective theories that the circular Crunch Berries were actually *berries*. Understandably, the court found that interpretation unreasonable. Many other cases are resolved with similarly direct results based on an interpretation that is actually *reasonable*.[18] For example, a 2015 case against Makers' Mark whiskey found that the plaintiff's interpretation of the term "handmade" was implausible under the reasonable consumer standard because "nobody could believe a bourbon marketed this widely at this volume is made entirely or predominantly by hand."[19] Relying on dictionary definitions, the court noted that bourbon "cannot be grown in the wild," like coffee or orange juice, and was in that sense

---

15. Erica A. Burgos, *Selling the Footlong Short: How Consumers Inch toward Satisfaction in Costly Food Class Action Litigation*, 13 SEVENTH CIRCUIT REV. 259, 266 (2017).

16. Timothy Williams, *Dannon Settles with F.T.C. over Some Health Claims*, N.Y. TIMES, Dec. 15, 2010, https://www.nytimes.com/2010/12/16/business/16yogurt.html.

17. *See* Sipos et al., *supra* note 5.

18. *See* Ang v. Whitewave Foods Co., No. 13-CV-1953, 2013 WL 6492353, at *4 (N.D. Cal. Dec. 10, 2013) ("It is simply implausible that a reasonable consumer would mistake a product like soymilk or almond milk with dairy milk from a cow . . . Under the Plaintiffs' logic, a reasonable consumer might also believe that veggie bacon contains pork, that flourless chocolate cake contains flour, or that e-books are made out of paper.").

19. Salters v. Beam Suntory, Inc., No. 4:14CV659-RH/CAS, 2015 WL 2124939, at *1 (N.D. Fla. May 1, 2015).

"handmade" as distinguished from the work of nature.[20] The court held that "no reasonable person would understand 'handmade' in this context to mean literally made by hand" without use of substantial equipment.

The "reasonable consumer" defense remains an important tool for defendants in food and beverage class actions. In 2018, courts relied on the defense in lawsuits that offered strained and implausible definitions of disputed labeling terms. Courts continued this trend in 2019, and several key dismissals were affirmed on appeal.[21]

What Is the Reasonable Consumer Defense and When Does It Come into Play?

Food litigation matters are generally putative consumer class actions alleging a false or misleading labeling claim. Plaintiffs generally use the vehicle of a state consumer protection statute, such as California's Unfair Competition Law,[22] in attempting to secure class-wide relief on behalf of consumers. New York and California are two of the most significant venues for such claims and historically have seen the majority of food litigation matters, with the Northern District of California earning the distinction as the nation's "food court," with even the court itself acknowledging the moniker.[23] The reason these venues tend to lead in filings is largely practical: it has become increasingly difficult to gain class certification of a 50-state nationwide consumer class action, principally because individual variation among state consumer protection laws make such cases unmanageable.[24] So, among single-state classes, California and New York, with their large urban populations and favorable jury pools, tend to lead the way.

A threshold question in cases alleging violations of state consumer protection law is often whether the labeling claims are likely to mislead

---

20. *Id.* at *2.
21. Perkins Coie, Food Litigation 2019 Year in Review 9 (2020), https://www.perkinscoie.com/images/content/2/2/229474/2019-Food-Litigation-YIR-v4.pdf.
22. Cal. Bus. & Prof. Code § 17200 *et seq.* (1977).
23. *See* Jones v. ConAgra Foods, Inc., No. C 12-01633 CRB, 2014 WL 2702726, at *1, n.1 (N.D. Cal. June 13, 2014) (acknowledging "the flood" of food litigation cases being filed in the Northern District of California and its accompanying reputation as the "food court").
24. *See* Mazza v. Am. Honda Motor Co., 666 F.3d 581, 596 (9th Cir. 2012) ("Because the law of multiple jurisdictions applies here to any nationwide class of purchasers . . . variances in state law overwhelm common issues and preclude predominance for a single nationwide class.").

the public, with courts asking whether "a significant portion of the general consuming public or of targeted consumers, acting reasonably in the circumstances, could be misled."[25] Under New York law, for example, in the consumer protection context, a statement is deceptive only if it is likely to mislead a reasonable consumer.[26] In making this determination, "context is crucial."[27] A case alleging "diet" soda would lead to weight loss was found implausible in large part because the plaintiffs were reading the word "diet" out of context. Unlike a "diet" pill, "diet" soda was referring to the type of soda, and that it was the "diet" variety of that soda, in that it had reduced or zero calories compared to other sodas of that brand.[28] Put simply, isolating single words or phrases on labeling is often insufficient to make out a plausible case given the reasonable consumer standard.

Of course, state courts have also contributed to the development on reasonable consumer jurisprudence, especially since plaintiffs bring these cases so often under state consumer protection statutes. One such case was that of *Brady v. Bayer Corp.*[29] *Brady*, assessing whether reasonable consumers would be misled when One A Day brand vitamins suggested a serving size of two pills, identified four distinct categories of reasonable consumer cases. These were:

- **Common sense**. "If a claim of misleading labeling runs counter to ordinary common sense or the obvious nature of the product, the claim is fit for disposition at the [motion to dismiss] stage of the litigation."[30]
- **Literal truth or falsity**. "Literal truth can sometimes protect a product manufacturer from a mislabeling claim, but it is no guarantee."[31]

---

25. Lavie v. Procter & Gamble Co., 105 Cal. App. 4th 496, 508 (Cal. Ct. App. 2003).
26. *See* Fink v. Time Warner Cable, 714 F.3d 739, 741 (2d Cir. 2013).
27. *Id.* at 742; Koenig v. Boulder Brands, Inc., 995 F. Supp. 2d 274, 288 (S.D.N.Y. 2014) ("[I]n resolving the reasonable consumer inquiry, one must consider the entire context of the label.") (quotation marks omitted). *See also* Nat'l Labor Relations Bd. v. Federbush Co., 121 F.2d 954, 957 (2d Cir. 1941) (Hand, J.) ("Words are not pebbles in alien juxtaposition; they have only a communal existence; and not only does the meaning of each interpenetrate the other, but all in their aggregate take their purport from the setting in which they are used.").
28. Manuel v. Pepsi-Cola Co., No. 17 CIV. 7955 (PAE), 2018 WL 2269247, at *8 (S.D.N.Y. May 17, 2018), *aff'd*, 763 F. App'x 108 (2d Cir. 2019).
29. 26 Cal. App. 5th 1156 (Cal. Ct. App. 2018).
30. *Id.* at 1165.
31. *Id.* at 1166.

- **The front-back dichotomy.** "[Q]ualifiers in the packaging can ameliorate any tendency of the label to mislead."[32]
- **Brand names being misleading in themselves.** "Brand names by themselves may be misleading in the context of the product being marketed."[33]

Generally, the reasonable consumer standard comes into play at the motion to dismiss stage, although it may also be an argument at class certification or summary judgment.[34] Pursuant to the Federal Rules of Civil Procedure, plaintiffs must make a short and plain statement of the claim, with heightened pleading requirements for allegations of fraud.[35] Under prevailing federal standards, these claims must be *plausible*, not simply possible.

At the motion to dismiss stage, courts must construe allegations in the light most favorable to the non-moving party, the plaintiff. But that does not mean the court must adopt the plaintiff's allegations wholesale. The reasonable consumer standard is an important check, even at this pleading stage. Indeed, it can be a dispositive defense. Under the reasonable consumer standard, a plaintiff demonstrates "more than the mere possibility that the label 'might conceivably be misunderstood by some few consumers viewing it in an unreasonable manner.'"[36] "Rather, the reasonable consumer standard requires a probability that a significant

---

32. *Id.* at 1167.

33. *Id.* at 1170 (noting also that this more often arose in the Lanham Act context).

34. The sufficiency of proof needed to sustain a claim under the reasonable consumer standard at the summary judgment stage is an issue that remains unsettled in federal courts. Although it is clear enough that a plaintiff must come forward with *some* proof of deception to make it to trial. *See* Ries v. Ariz. Beverages USA LLC, No. 10-01139 RS, 2013 WL 1287416 (N.D. Cal. Mar. 28, 2013) ("[Plaintiffs] have neither intrinsic evidence that the labels are false (because [high-fructose corn syrup] and citric acid are not natural) or extrinsic evidence that a significant portion of the consuming public would be confused by them.") (granting summary judgment under reasonable consumer standard for failure of proof). *See also* Beardsall v. CVS Pharmacy, Inc., 953 F.3d 969, 976 (7th Cir. 2020) ("This is not to say that extrinsic evidence in the form of consumer surveys or market research is *always* needed for a plaintiff to survive summary judgment [but] such evidence is necessary where the advertising is not clearly misleading on its face and materiality is in doubt.") (affirming summary judgment for defendant where plaintiff failed to provide sufficient evidence to show aloe vera gel product labels misled consumers by not disclosing low concentrations of a compound called acemannan found in aloe vera) (emphasis in original).

35. *See* Fed. R. Civ. P. 8, 9(b).

36. Ebner v. Fresh, Inc., 838 F.3d 958, 965 (9th Cir. 2016) (quoting Lavie v. Procter & Gamble Co., 105 Cal. App. 4th 496 (Cal. Ct. App. 2003)).

portion of the general consuming public or of targeted consumers, acting reasonably in the circumstances, could be misled."

Generally, questions of fact are not appropriate for resolution at the motion to dismiss stage.[37] Even so, courts recognize that they may make determinations on the reasonable consumer standard based on the product packaging at issue even at this stage.[38] As such, the reasonable consumer standard is an important check on the plausibility of a plaintiff's claim that is *addressable* by a court at the motion to dismiss phase.

If disputes regarding the reasonable consumer standard are not resolved at the Rule 12 stage, the case might proceed to class certification or summary judgment. At these stages, aided by discovery, each side might provide expert testimony that a reasonable consumer would or would not be misled by the challenged labeling and attempt to address any fact issues identified by the court preventing adjudication on reasonable consumer grounds at the motion to dismiss stage. The plaintiff's anecdotal evidence would be insufficient, so courts require some indication that reasonable consumers are misled by the challenged claims to defeat a reasonable consumer defense.

Ongoing Debates in Reasonable Consumer Jurisprudence

Just who is the "reasonable consumer?" Of course, it is an *objective* standard and courts assess reasonableness based on context, based on the circumstances of the case. Reading words in context, for example, could help distinguish between *Crunch Berries* and *Berries* that go *Crunch*.

What courts require of reasonable consumers is a matter of some debate between courts. Key questions remain unresolved at a national level, such as whether a reasonable consumer reads the back of the box for ingredient information or other contextual information for the challenged claim. Even though many courts would say yes, some important jurisdictions disagree. With so many cases pending in federal courts across the country, the reasonable consumer standard is constantly developing.[39]

---

37. Williams v. Gerber Prods. Co., 552 F.3d 934, 938 (9th Cir. 2008) (noting whether a business practice is deceptive will usually be a question of fact not appropriate for determination at the pleadings stage).

38. Moreno v. Vi-jon, Inc., No. 20CV1446 JM(BGS), 2021 WL 807683, at *5 (S.D. Cal. Mar. 3, 2021) (collecting cases).

39. Some argue that the FTC should step in and update their guidance to refine what factors a reasonable consumer relies upon when making their purchase decisions. Burgos, *supra* note 15, at 280.

The Reasonable Consumer Standard Is an Objective Test

Under the reasonable consumer standard, it is not sufficient for a plaintiff to demonstrate her subjective belief about a food product.[40] She could, sincerely but erroneously, believe in her novel interpretation of a product's labeling claims. That subjective belief is not actionable, unless plaintiffs can show a probability that a significant portion of consumers acting reasonably would be misled. The reasonable consumer is the hypothetical representative of "a significant portion of the general consuming public or of targeted consumers."[41] Because the reasonable consumer represents the public at large, "the experience of very few persons or isolated examples may not suffice to show false or misleading statements."[42]

Just How Sophisticated Is the Reasonable Consumer?

Under an objective standard, the reasonable consumer need not be "exceptionally acute and sophisticated," but the reasonable consumer must be more than the "least sophisticated consumer."[43] In other words, the reasonable consumer is "neither the most vigilant and suspicious" nor "the most unwary and unsophisticated," but instead is "the ordinary consumer" within the targeted population.[44] A reasonable consumer is not an "unwary consumer."[45] "Simply put, a reasonable consumer does not check her common sense at the door of the store."[46]

---

40. Suchanek v. Sturm Foods, Inc., 764 F.3d 750, 758 (7th Cir. 2014) (reasonable consumer standard not based on consumer's "subjective understanding of a package"); *accord* Hughes v. Ester C Co., 330 F. Supp. 3d 862, 871 (E.D.N.Y. 2018) ("[I]t is not enough for a plaintiff to assert, based on his or her own subjective belief that a statement on the defendant's label conveyed the alleged implied message.").

41. *Ebner*, 838 F.3d at 965; Beardsall v. CVS Pharmacy, Inc., 953 F.3d 969, 973 (7th Cir. 2020) (citing *Ebner*); Jessani v. Monini N. Am., Inc., 744 F. App'x 18, 19 (2d Cir. 2018) (same).

42. Tubbs v. AdvoCare Int'l, L.P., 785 F. App'x 396, 397 (9th Cir. 2019) (citation and internal quotation marks omitted).

43. Brod v. Sioux Honey Ass'n, Coop., 927 F. Supp. 2d 811, 828 (N.D. Cal. 2013), *aff'd*, 609 F. App'x 415 (9th Cir. 2015); Becerra v. Dr Pepper/Seven Up, Inc., 945 F.3d 1225, 1228 (9th Cir. 2019) (rejecting the "least reasonable consumer" standard).

44. Utts v. Bristol-Myers Squibb Co., 251 F. Supp. 3d 644, 683 (S.D.N.Y. 2017), *aff'd sub nom.* Gibbons v. Bristol-Myers Squibb Co., 919 F.3d 699 (2d Cir. 2019).

45. Freeman v. Time, Inc., 68 F.3d 285, 289 (9th Cir. 1995) (rejecting "unwary consumer" standard); Haskell v. Time, Inc., 857 F. Supp. 1392, 1398 (E.D. Cal. 1994) (same).

46. Weiss v. Trader Joe's, No. 19-55841, 2021 WL 816075, at *1 (9th Cir. Mar. 3, 2021); *accord* Daniel v. Mondelez Int'l, Inc., 287 F. Supp. 3d 177, 193 (E.D.N.Y. 2018) ("A reasonable consumer does not lack common sense.").

Judges are consumers as well, who themselves may have views on what would be reasonable in how to read product labels.[47] Given that the analysis of the reasonable consumer standard is often dispositive in food litigation matters, courts generally devote significant portions of their analysis to wrestling with whether a plaintiff's interpretation would likely mislead a reasonable consumer. This analysis may rely on common sense, experience, sophistication of the target consumer, or other factors. Below are example cases where the reasonable consumer standard led a court to dismiss a claim.

- A reasonable consumer expects that there to be ice in iced drinks.[48]
- A reasonable consumer expects that a packaged orange juice product might not be "fresh squeezed."[49]
- A reasonable consumer expects that products labeled soy milk and almond milk are not dairy milk.[50]
- A reasonable consumer expects that bourbon may be made using machines, even if it is labeled "handcrafted" or "handmade."[51]

Courts Rely on a Number of Factors in Determining Reasonableness

The reasonable consumer "act[s] reasonably under the circumstances."[52] To determine what is reasonable, judges rely on a variety of factors, including but not limited to those listed below.[53]

Actual Truth of the Challenged Claim

If an advertising statement is factually true, courts are quick to rely on that when determining whether a reasonable consumer would be misled. This is because "literal truth can sometimes protect a product manufacturer from a misleading claim, but it is no guarantee."[54] As one court put

---

47. *See* Dumont v. Reily Foods Co., 934 F.3d 35, 41 (1st Cir. 2019) ("[W]e [i.e., judges] ourselves would likely land upon that reading [of the label] were we in the grocery aisle with some time to peruse the package.").
48. Galanis v. Starbucks Corp., No. 16 C 4705, 2016 WL 6037962, at *3 (N.D. Ill. Oct. 14, 2016).
49. Veal v. Citrus World, Inc., No. 2:12-CV-801-IPJ, 2013 WL 120761, at *4 n.4 (N.D. Ala. Jan. 8, 2013).
50. Ang v. Whitewave Foods Co., No. 13-CV-1953, 2013 WL 6492353, at *4 (N.D. Cal. Dec. 10, 2013).
51. Welk v. Beam Suntory Imp. Co., 124 F. Supp. 3d 1039, 1044 (S.D. Cal. 2015).
52. Ebner v. Fresh, Inc., 838 F.3d 958, 965 (9th Cir. 2016).
53. FTC Policy Statement on Deception, *supra* note 11 (a "trier of fact" will consider "many factors in determining the reaction of the ordinary consumer to a claim or practice").
54. Moore v. Mars Petcare US, Inc., 966 F.3d 1007 (9th Cir. 2020).

it, the truth of the claim is "a common starting point."[55] In *Cheslow*, a case alleging that a chocolate company's white baking chips were misrepresented as white chocolate, "both defendant and plaintiffs agree[d]" that there was "no affirmative statement or representation on the product [of white baking chips]" that was "false or deceptive."[56] "For example, the package label does not include the word 'chocolate' or 'cocoa.'"[57] The test under such circumstances, then, is whether the plaintiff can demonstrate that "the product, despite containing only true statements, has the capacity, likelihood, or tendency to deceive the consuming public."[58]

Putting the test to practice, several courts find that factually true statements are not misleading. In *Ebner v. Fresh, Inc.*, the plaintiff's claim that "the reasonable consumer would be deceived as to the amount of lip product in a tube of [lip balm was] not plausible" because it "was undisputed that the [lip balm's] label disclose[d] the correct weight of included lip product."[59] Similarly, reasonable consumers would not interpret a claim of "fat-free," which was objectively true, to also communicate that the product was low in sodium.[60]

Looking at Context

Courts often look to the context of the challenged claim, including the presence of any disclaimers or clarifying language, especially if such language is adjacent to the allegedly misleading statements. As the Second Circuit put it, "context is crucial," and the Ninth Circuit agreed that "context of the packaging as a whole" must be considered in evaluating whether deception has occurred.[61] This is a particularly hot debate in reasonable consumer law.

A disclaimer *alone* might not cure a potentially misleading statement. For example, in *Hughes v. Ester C Co.*, the court denied dismissal of claims against a vitamin manufacturer, noting:

> At this early stage of the litigation, it cannot be determined whether a disclaimer on the back of . . . [the] products, stating that it is "not intended to diagnose, treat, or prevent any disease," eliminates the possibility of a reasonable consumer being misled

---

55. Cheslow v. Ghirardelli Chocolate Co., 445 F. Supp. 3d 8, 16 (N.D. Cal. 2020).
56. *Id.*
57. *Id.*
58. *Id.*
59. 838 F.3d 958, 965 (9th Cir. 2016).
60. Figy v. Frito-Lay N. Am., Inc., 67 F. Supp. 3d 1075, 1091 (N.D. Cal. 2014).
61. Fink v. Time Warner Cable, 714 F.3d 739, 742 (2d Cir. 2013); Williams v. Gerber Prods. Co., 552 F.3d 934, 939 n.3 (9th Cir. 2008); Belfiore v. Procter & Gamble Co., 311 F.R.D. 29, 53 (E.D.N.Y. 2015) (the "entire mosaic [is] viewed rather than each tile separately").

into thinking . . . [the allegedly misleading statements] signified a cold or flu prevention product.[62]

Font size and emphasis compared to the allegedly misleading statement may be factors that courts consider.[63] For example, a leading case in the Ninth Circuit on reasonable consumer law, *Freeman v. Time, Inc.*, regarded mail-in sweepstakes. The appellate court noted that the qualifying language there was not "unreasonably small" and "no reasonable consumer could ignore it."[64]

In addition to font size and emphasis, courts might consider the clarifying language's placement on the package. For example, putting "vegan" next to "butter" or "veggie" next to "burger" could change the meaning to a reasonable consumer.[65] Proximity to the allegedly misleading statement can dispel allegations of deception.

But when this qualifying language appears on the side or back of the package, courts are split. Some note that "qualifiers in packaging, usually on the back of a label or in ingredient lists, can ameliorate any tendency of the label to mislead."[66] Others do not expect that reasonable consumers should consult nutrition facts panels or ingredient lists, especially if it contradicts the front panel's language.[67] Still other courts note that where there

---

62. Hughes v. Ester C Co., 930 F. Supp. 2d 439, 463–65 (E.D.N.Y. 2013).

63. *Compare* Kommer v. Bayer Consumer Health, 252 F. Supp. 3d 304, 312 (S.D.N.Y. 2017) (no reasonable consumer misled where a "disclaimer [was] printed in reasonably-sized font right at the top of the Instructions" that were allegedly deceptive), *with* Koenig v. Boulder Brands, Inc., 995 F. Supp. 2d 274, 287–88 (S.D.N.Y. 2014) ("[A] reasonable consumer might . . . focus on the more prominent portion of the product label that touts the product as 'Fat Free Milk and Omega-3s,' and overlook the smaller text that discloses the fat content on the front of the carton or the nutrition label."), *and* Delgado v. Ocwen Loan Servicing, LLC, No. 13-CV-4427, 2014 WL 4773991, at *9 (E.D.N.Y. Sept. 24, 2014) (holding that disclaimers were "not conspicuous or prominent enough to necessarily cure" alleged misrepresentation).

64. Freeman v. Time, Inc., 68 F.3d 285, 289 (9th Cir. 1995).

65. *See* Order Granting in Part and Denying in Part Motion for Preliminary Injunctive Relief at 7, Miyoko's Kitchen v. Ross, No. 3:20CV00893 (N.D. Cal. Aug. 21, 2020) (manufacturer's use of the word "butter," in immediate or close proximity to terms like "vegan," "made from plants," "cashew cream fermented with live cultures," and "cashew & coconut oil spread," was not misleading); Turtle Island Foods SPC v. Foman, 424 F. Supp. 3d 552, 574 (E.D. Ark. 2019) (label of plant-based "meat" products "include[d] ample terminology to indicate [their] vegan or vegetarian nature").

66. Moore v. Mars Petcare US, Inc., 966 F.3d 1007 (9th Cir. 2020).

67. *See* Mantikas v. Kellogg Co., 910 F.3d 633, 637 (2d Cir. 2018) ("We conclude that a reasonable consumer should not be expected to consult the Nutrition Facts panel on the side of the box to correct misleading information set forth in large bold type on the front of the box."). Bell v. Publix Super Markets, Inc., 982 F.3d 468, 477 (7th Cir. 2020) ("[O]ur colleagues in other circuits held that the reasonable consumer

is any ambiguity from the front panel that is confirmed or clarified by the back or side panel's language, it is relevant to a reasonable consumer.[68]

Use of Dictionary Definitions

When a case turns on a particular term, courts often look to dictionary definitions. For example, in *Becerra v. Dr Pepper/Seven Up, Inc.*, a "diet" soda case, the Ninth Circuit relied upon Merriam-Webster's definition of "diet" to debunk the plaintiff's theory.[69] In the Second Circuit, the dictionary definition of "steak" was used when a consumer complained that his "steak" was actually ground beef.[70] Noting that Merriam-Webster defined "steak" in relevant part as "ground beef prepared for cooking or serving in the manner of a steak," the appellate panel affirmed dismissal on reasonable consumer grounds.[71]

Compliance with State or Federal Labeling Standards

Courts might also rely on whether the defendant has complied with state or federal labeling standards as suggestive that a reasonable consumer would not have been misled. Compliance suggests a reasonable consumer would not be misled; however, the factor alone is not dispositive.[72] California, in particular, provides a "safe harbor" for defendants sued under California's False Advertising Law and Unfair Competition Law where

---

standard does not presume, at least as a matter of law, that reasonable consumers will test prominent front-label claims by examining the fine print on the back label. We agree with that reasoning.").

68. *See, e.g.*, Moreno v. Vi-jon, Inc., No. 20CV1446 JM(BGS), 2021 WL 807683, at *6 (S.D. Cal. Mar. 3, 2021) ("Defendant's use of the word germ is clarified by the disclosure on the back panel, namely that the hand sanitizer is effective at eliminating 99.99% of many *common* harmful germs and bacteria"); Workman v. Plum Inc., 141 F. Supp. 3d 1032, 1035 (N.D. Cal. 2015) (holding that, where "a reasonable consumer would simply not . . . assume that the size of the items pictured [on the front of a package] directly correlated with their predominance [in the product]," and where "any potential ambiguity could be resolved by the back panel of the products," the packaging was not deceptive); Solak v. Hain Celestial Group, Inc., No. 317CV0704LE-KDEP, 2018 WL 1870474, at *9 (N.D.N.Y. Apr. 17, 2018) ("[T]o the extent that the Packaging [of defendant's veggie straws] does include arguably ambiguous representations (e.g., the 'smart and wholesome' language, the 'garden grown potatoes' and 'ripe vegetables' language, etc.), those representations are adequately clarified by the other statements and images included on the Packaging.").

69. 945 F.3d 1225, 1229 (9th Cir. 2019).

70. Chen v. Dunkin' Brands, 954 F.3d 492, 500–01 (2d Cir. 2020).

71. *Id.*

72. Blue Buffalo Co. v. Nestle Purina Petcare Co., No. 4:15 CV 384 RWS, 2015 WL 3645262, at *5 (E.D. Mo. June 10, 2015) ("While compliance with regulatory authority may be relevant to Purina's defense, it does not preclude liability for implied false advertising as a matter of law.").

"the Legislature has permitted certain conduct or considered a situation and concluded no action shall lie."[73] And while alleged noncompliance with some federal regulatory standard might give rise to a claim that the label is "unlawful" under state consumer protection laws, the bare reference to some alleged technical regulatory violation is not generally considered sufficient to plead that a "reasonable consumer" has been misled.[74]

Use of Consumer Surveys

As a recent trend, plaintiffs have been using consumer surveys in the pleadings in an attempt to demonstrate that a substantial amount of the public might be misled by the challenged claim. Many courts have been skeptical of these surveys (and rightly so) as the surveys themselves may be leading and suggestive.[75] Further, these surveys may be so general as to be insufficient to "bolster [the] plaintiff's claim into the realm of plausibility."[76]

Puffery

Reasonable consumers do not rely on puffery because a reasonable consumer knows "puffery is a fact of life."[77] Statements are "mere puffery" and therefore nonactionable if they "cannot be proven either true or false,"[78] or are too "generalized, vague, and unspecif[ied]" to communicate any "objective and measurable representation about the product."[79]

---

73. Barber v. Nestle USA, Inc., 154 F. Supp. 3d 954, 962 (C.D. Cal. 2015) ("Plaintiffs may wish—understandably—that the Legislature had required disclosures beyond the minimal ones required by [the relevant statute]. But that is precisely the sort of legislative second-guessing that the safe harbor doctrine guards against.").

74. Twohig v. Shop-Rite Supermarkets, Inc., No. 20-CV-763 (CS), 2021 WL 518021, at *6 (S.D.N.Y. Feb. 11, 2021) (rejecting argument that alleged violation of labeling requirements provide proof of deception) ("There is no extrinsic evidence that the perceptions of ordinary consumers align with [U.S. Food and Drug Administration's] various labeling standards.").

75. Strumlauf v. Starbucks Corp., No. 16-CV-01306-YGR, 2018 WL 306715, at *7 (N.D. Cal. Jan. 5, 2018) (disregarding two consumer surveys on a motion for summary judgment because the questions were "leading and suggestive"); Cheslow v. Ghirardelli Chocolate Co., No. 19-CV-07467-PJH, 2020 WL 4039365 (N.D. Cal. July 17, 2020) (finding plaintiffs, who relied on a survey where nearly 92 percent of respondents who viewed front panel of chocolate manufacturer's packaging for premium white baking chips indicated that they thought it contained white chocolate, failed to allege that a reasonable consumer would have been deceived by packaging because survey only showed respondents front panel of packaging, and by omitting back panel, survey deprived respondents of relevant information, namely, the ingredient list).

76. Yu v. Dr Pepper Snapple Group, No. 18-cv-06664-BLF, 2020 WL 5910071 (N.D. Cal. Oct. 6, 2020).

77. Wysong Corp. v. APN, Inc., 889 F.3d 267, 271 (6th Cir. 2018).

78. Time Warner Cable, Inc. v. DirecTV, Inc., 497 F.3d 144, 160 (2d Cir. 2007).

79. Glen Holly Entm't, Inc. v. Tektronix Inc., 343 F.3d 1000, 1015 (9th Cir. 2003).

"[T]o be actionable as an affirmative misrepresentation, a statement must make a specific and measurable claim, capable of being proved false or of being reasonably interpreted as a statement of objective fact."[80]

Some examples of puffery include:

- Statements that veggie straw products were a "smart and wholesome" way of satisfying consumers' "snack craving[s]"[81]
- The word "refresh" and the "hundreds of plus symbols" on the label of a specialty water[82]
- The word "premium" in the phrase "premium baking chips"[83]

Conclusion

The reasonable consumer defense is here to stay. It is an important defense in consumer class action cases generally and specifically in the food litigation context against frivolous cases and dubious theories before discovery. For example, the courts have seen more than 100 cases filed on virtually identical theories on vanilla flavoring since 2019.[84] Federal courts in New York have seen the lion's share of these cases, and already many of these courts have dismissed these claims on reasonable consumer grounds.[85] As the Southern District of New York noted, "When consumers read vanilla on a product label, they understand it to mean the product has a certain taste. It is difficult to comprehend what is misleading when the Defendant's 'Smooth Vanilla' [product] tastes like vanilla."[86] Whether its accusations that vanilla ice cream is not made with real vanilla, that "handmade" bourbon is not made by hand, or that Crunch Berries are not made with real berries, the reasonable consumer defense is here to stay.

---

80. Vitt v. Apple Computer, Inc., 469 F. App'x 605, 607 (9th Cir. 2012) (internal quotation marks omitted).

81. Solak v. Hain Celestial Group, Inc., No. 317CV0704LEKDEP, 2018 WL 1870474, at *6 (N.D.N.Y. Apr. 17, 2018).

82. Weiss v. Trader Joe's Co., No. 818CV01130JLSGJS, 2018 WL 6340758, at *4 (C.D. Cal. Nov. 20, 2018), aff'd sub nom. Weiss v. Trader Joe's, No. 19-55841, 2021 WL 816075 (9th Cir. Mar. 3, 2021).

83. Cheslow v. Ghirardelli Chocolate Co., 445 F. Supp. 3d 8, 18 (N.D. Cal. 2020) ("There is no obvious way to measure 'premium.'").

84. PERKINS COIE, supra note 14, at 9.

85. See Steele v. Wegmans Food Mkts., Inc., 472 F. Supp. 3d 47, 50 (S.D.N.Y. 2020); Pichardo v. Only What You Need, Inc., No. 20-CV-493 (VEC), 2020 WL 6323775, at *5 (S.D.N.Y. Oct. 27, 2020); Cosgrove v. Blue Diamond Growers, No. 19 Civ. 8993 (VM), 2020 WL 7211218, at *4 (S.D.N.Y. Dec. 7, 2020); Barreto v. Westbrae Natural, Inc., No. 19-cv-9677 (PKC), 2021 WL 76331, at *4 (S.D.N.Y. Jan. 7, 2021).

86. Pichardo, 2020 WL 6323775, at *5.

Chapter 4

Food Safety

James F. Neale and Benjamin P. Abel

This chapter will provide a basic overview of food safety, including the regulatory environment governing food companies, as well as various issues and concerns pressing in the industry.

Introduction to the Regulatory Framework

A unique and defining aspect of the regulatory scheme overseeing the American food supply is the absence of any one agency with food safety responsibility. Instead, the current regulatory scheme is fragmented, decentralized, and sometimes politicized. According to the U.S. Government Accountability Office, 15 federal agencies administer at least 30 acts related to food safety.[1] Although two of those agencies, the U.S. Food and Drug Administration (FDA) and U.S. Department of Agriculture (USDA), bear most responsibility for protecting and regulating America's food supply,[2] significant fragmentation and inefficiencies are trademarks of the current system. Most retail consumers have direct contact with the food supply at supermarkets and restaurants, facilities at which retail sale exemptions put most regulatory oversight into the hands of state and local

---

1. GEOFFREY S. BECKER & DONNA PORTER, CONGRESSIONAL RESEARCH SERVICE, THE FEDERAL FOOD SAFETY SYSTEM: A PRIMER 2 (2007).
2. *Id.*

health inspectors.[3] In recent years, calls for a more unified approach have increased, but have yet to be put into action.[4]

The Major Food Regulators at the Federal Level
USDA

USDA's responsibilities include keeping the United States' meat, poultry, and egg products safe for human consumption. USDA's approach to food safety is multilayered and best summarized by its goal of "enhancing food safety by taking steps to reduce the prevalence of foodborne hazards from farm to table" in those foods that it regulates.[5] The specific USDA entity primarily charged with enforcement of food safety standards is its Food Safety and Inspection Service (FSIS).

FSIS[6] is the USDA agency responsible for ensuring that all meat, poultry, and egg products entering the United States market are inspected, safe, wholesome, properly labeled, and properly packaged.[7] The agency coordinates its efforts with FDA, as well as a number of other coordinate governmental agencies sharing responsibility for food safety.[8] FSIS enforces the Federal Meat Inspection Act (FMIA), the Poultry Products Inspection Act (PPIA), and parts of the Egg Products Inspection Act (EPIA), which taken together dictate federal inspection and regulation of meat, poultry, and egg products prepared for human consumption in the United States.[9]

---

3. *See* 9 C.F.R. § 303.1 (2021) (generally exempting facilities engaged in business-to-consumer sales of meat and poultry from USDA inspection and regulation); 21 C.F.R. § 1.2227 (2021) (generally exempting facilities' business-to-consumer sales of other food products from FDA inspection and regulation).

4. *See* Dan Flynn, *Trump Wants a Single Federal Food Safety Agency Put under USDA*, Food Safety News, June 22, 2018, https://www.foodsafetynews.com/2018/06/president-trump-wants-the-single-federal-food-safety-agency-put-under-usda/.

5. Memorandum of Understanding between USDA and FDA concerning Information Sharing Related to Food Safety, Public Health, and Other Food-Associated Activities, https://www.fsis.usda.gov/sites/default/files/media_file/2020-07/USDA_FDA_Info_Sharing_MOU.pdf.

6. FSIS was so designated in 1981. *See* USDA FSIS, *Our History*, https://www.fsis.usda.gov/about-fsis/history (last visited May 4, 2021).

7. *See id.*

8. USDA, FY 2021 Budget Summary 55, *available at* https://www.usda.gov/sites/default/files/documents/usda-fy2021-budget-summary.pdf.

9. *See* USDA FSIS, *About FSIS*, https://www.fsis.usda.gov/about-fsis (last visited May 4, 2021).

The first of those acts, FMIA,[10] has a stated goal to protect consumers' health and welfare "by assuring that meat and meat food products distributed to them are wholesome, not adulterated, and properly marked, labeled, and packaged."[11] In addition to requiring that USDA ensure compliance with the Humane Methods of Livestock Slaughter Act, FMIA requires USDA to inspect all "amenable species"[12] before entry into slaughtering and packing facilities and to further ensure the separate treatment of any animals showing signs of disease following inspection.[13] Post-slaughter, FMIA requires that FSIS inspectors inspect the carcasses of animals and mark those that are unadulterated "Inspected and passed," while adulterated carcasses are tagged as "Inspected and condemned" and must be destroyed in the presence of the inspector.[14] The constant presence of USDA inspectors during operations at establishments under USDA authority is a defining characteristic of the regulatory scheme implemented under FMIA, which differs dramatically from facilities subject to infrequent inspection by FDA.

In addition to requiring pre- and post-slaughter inspections, FMIA directs USDA personnel to observe all subsequent packaging and to pre-approve all labeling. This labeling pre-approval requirement also differs dramatically from food products regulated by FDA, which does not approve labels on the food products it regulates. USDA-approved labels must be legible and contain specified information, such as identifying information for the packer and distributor and an accurate statement of the contents and their derivation.[15] In addition to establishing its extensive inspection regime, FMIA outlaws the introduction into commerce of food products manufactured inconsistently with its provisions.[16] Establishments subject to the act are obligated to have a recall procedure in place and to produce substantiating documentation of their processes to USDA inspectors upon request.[17]

---

10. 21 U.S.C. § 601 *et seq.*
11. *Id.* § 602 (1967).
12. Cattle, sheep, swine, goats, horses, mules, other equines, and any additional species of livestock that the secretary of agriculture considers appropriate including, perhaps curiously, imported catfish. *See* 21 U.S.C. § 601(w); Geoffrey S. Becker, Congressional Research Service, Meat and Poultry Inspection: Background and Selected Issues 1 (2010), *available at* http://nationalaglawcenter.org/wp-content/uploads/assets/crs/RL32922.pdf.
13. *See* 21 U.S.C. § 603.
14. *Id.* § 604.
15. *Id.* §§ 607(a)–(b), 601(n), 619.
16. *Id.* §§ 609–613.
17. *Id.* § 614.

USDA is also responsible for enforcing PPIA, which is similar to FMIA in many respects.[18] PPIA's mission is to assure that poultry products distributed within the United States for human consumption "are wholesome, not adulterated, and properly marked, labeled, and packaged."[19] PPIA centers on its section 455 inspection program, which requires pre-mortem inspections of poultry, defined as any domesticated bird, live or dead,[20] in establishments processing poultry or poultry products intended for interstate commerce.[21] Section 455 further requires post-mortem inspection of processed birds and, when appropriate, quarantining, segregating, and reinspecting poultry and poultry products.[22]

The act defines adulterated poultry and poultry products as those containing poisonous or deleterious substances, including pathogens and color additives,[23] and poultry having been intentionally subjected to radiation.[24] PPIA requires that establishments adhere to sanitary practices established by USDA, and it provides for the removal of inspection services for noncompliance with the act.[25] In addition to mandating FSIS's comprehensive inspection program, PPIA, like FMIA, regulates pre-approved labeling of products under its authority.

Finally, USDA is partially responsible for enforcing EPIA, which mandates that egg products are "wholesome, otherwise not adulterated, and properly labeled and packaged."[26] EPIA imposes inspection requirements for two categories of eggs—egg products and shell eggs.[27] It gives enforcement authority to both USDA and FDA. The act directs USDA to continuously inspect operations at plants and facilities that break and process shell eggs into dried, frozen, or liquid egg products, while FDA is responsible for periodic inspection of egg substitutes, imitation eggs, and similar products.[28] However, under a recent proposed rule from USDA,

---

18. *See* USDA FSIS, *supra* note 9.
19. 21 U.S.C. §§ 451–452.
20. *Id.* § 453(e).
21. *Id.* § 455(a).
22. *Id.* § 455(b).
23. *Id.* § 453(g)(1)–(2).
24. *Id.* § 453(g)(7).
25. *Id.* §§ 456, 467.
26. *Id.* § 1031.
27. *See* USDA FSIS, *Egg Products Inspection Act*, https://www.fsis.usda.gov/policy/food-safety-acts/egg-products-inspection-act (last updated Jan. 21, 2016).
28. 21 U.S.C. § 1034(a); USDA FSIS, *supra* note 27.

FSIS would take authority over egg substitutes and freeze-dried egg products.[29]

EPIA prohibits trafficking in "restricted eggs," defined as eggs with broken or cracked shells, unbroken but dirty eggs, "incubator rejects," and eggs that are sour, moldy, or otherwise classified as inedible, or the use of such eggs in the preparation of egg products.[30] The act additionally requires the pasteurization of egg products and refrigeration of eggs during storage and transport.[31] Like the other USDA regulatory schemes, EPIA also requires the labeling of egg products with the official inspection legend and number of the processing plant, and it prohibits false or misleading labels and containers.[32]

FDA

With the general exceptions of meat, poultry, and egg products,[33] FDA is responsible for ensuring the safety and accurate labeling of nearly all remaining food products in the United States, including produce, dairy products, seafood, and processed foods.[34] FDA is part of the U.S. Department of Health and Human Services and comprises regional field offices and centers, including the Office of Regulatory Affairs, which leads all FDA field activities,[35] and the Center for Biologics Evaluation and Research.[36] The FDA entity with the largest responsibility for food safety is the Center for Food Safety and Applied Nutrition (CFSAN).

CFSAN's mission is to both protect and promote public health "by ensuring that the nation's food supply is safe, sanitary, wholesome, and honestly labeled, and that cosmetic products are safe and properly labeled."[37] Along with its oversight of facilities that manufacture, process, pack, or hold foods regulated by FDA, CFSAN also provides support and technical assistance to restaurants, retailers, and other establishments

---

29. *See* Riëtte van Laack, *FSIS Proposed Egg Products HACCP Rule That Expands Jurisdiction*, FDA L. BLOG, Feb. 19, 2018, https://www.fdalawblog.net/2018/02/fsis-proposes-egg-products-haccp-rule-that-expands-jurisdiction/.
30. 21 U.S.C. §§ 1037(a) 1033(g), 1046(a)(1).
31. *Id.* §§ 1036(a), 1034(e), 1037(c).
32. *Id.* § 1036(a)–(b). Appeals of such determinations may be made to the U.S. court of appeals in which the facility in question is located.
33. *See* FDA, *What We Do*, http://www.fda.gov/aboutfda/whatwedo/default.htm (last updated Mar. 28, 2018).
34. BECKER & PORTER, *supra* note 1, at 2.
35. *See* FDA, *FDA Organization Charts*, http://www.fda.gov/AboutFDA/CentersOffices/OrganizationCharts/default.htm (last updated Mar. 21, 2019).
36. *See id.*
37. *See* FDA, *supra* note 33.

primarily regulated by state or local authorities.[38] In addition, CFSAN conducts food safety research, regulates food additives and bioengineered ingredients, coordinates FDA's surveillance and compliance programs, and develops and publishes food safety and regulatory information.[39]

The vast majority of FDA's authority with respect to human food and food safety, comes from the Federal Food, Drug, and Cosmetic Act of 1938 (FDCA). FDCA authorizes FDA to monitor the safety and accuracy of labeling for a wealth of products, including prescription and over-the-counter drugs, cosmetics, medical devices, blood and tissue products, animal drugs and feed, and the nation's food supply (with the exception of those products regulated by USDA).[40] Broadly speaking, FDCA prohibits the adulteration or misbranding of any food[41] to be introduced into interstate commerce.[42] In some cases, and under the authority of FDCA, FDA can prohibit interstate commerce involving certain foods entirely, such as raw milk.[43] The act also regulates labeling by defining any food whose label is false or misleading in any particular way as misbranded. Unlike USDA's FSIS, however, FDA does not pre-approve labels. FDCA further requires the prominent disclosure of information like the presence of artificial coloring, chemical preservatives, and pesticides.[44] As amended by the Nutrition Labeling and Education Act of 1990, FDCA also mandates certain nutritional information on product labels, including the caloric content, serving size, and the amounts of certain nutrients required by the act.[45] FDCA further prohibits "the alteration, mutilation, destruction, obliteration, or removal of the whole or any part of the labeling of" a food that results in the article being adulterated or misbranded.[46]

FDCA primarily attempts to guarantee food safety of products within FDA's jurisdiction through regulations requiring the industry to

---

38. *Id.*
39. *See id.*; BECKER & PORTER, *supra* note 1, at 3.
40. 21 U.S.C. § 301 *et seq.*
41. Defined by the act as "(1) articles used for food or drink for man or other animals, (2) chewing gum, and (3) articles used for components of any such article." *Id.* § 321(f). It includes vitamins and other dietary supplements intended for ingestion in tablet, capsule, or similar form. *See id.* §§ 343-2, 350, 350b.
42. *Id.* § 331(a)–(c), (g).
43. *See* 21 C.F.R. § 1240.61 (2021) (prohibiting sale of raw milk across state lines, even if legal in both the exporting and importing states). *See also* Organic Pastures Dairy Co., LLC v. Sebelius, No. 1:12-cv-02019-SAB, 2013 U.S. Dist. LEXIS 124121 (E.D. Cal. Aug. 29, 2013).
44. 21 U.S.C. § 343.
45. *Id.* § 343(q).
46. *Id.* § 331(k).

self-police to prevent adulteration and misbranding, through occasional inspections of manufacturing facilities and products already on the market, as well as through FDA exercising its seizure authority, when necessary.[47] Regulations provide that FDA agents may, within normal business hours, and upon written notice and the presentation of identification, enter any factory, warehouse, establishment, or vehicle in which food, drugs, devices, tobacco products, or cosmetics in or intended for interstate commerce are manufactured, processed, packed, held, or transported. Unlike USDA inspectors, however, FDA inspectors are not required to be present whenever a facility is operating and may only inspect a regulated facility once every several years. FDA agents are authorized to not only inspect, but they may also take samples as needed.[48] Inspection notices need not be given in advance, and FDA need not provide the reasons for the inspection.[49]

The most recent and important addition to the regulatory web governing food safety came in 2011 with the FDA Food Safety Modernization Act (FSMA). In fact, FSMA represents the most substantial food safety legislation enacted in many decades and constituted a fundamental paradigm shift towards FDA preventing outbreaks rather than simply reacting after they occur. FSMA imposed on FDA-regulated facilities requirements similar to those previously imposed on USDA-regulated enterprises. FSMA requires FDA-regulated facilities to register with FDA and pay fees, which FDA will use to fund more inspectors to conduct inspections of regulated facilities. The registration requirement also provides FDA with the ability to suspend the registration of any problem facility, essentially allowing it to summarily prohibit production of food at facilities it deems unsatisfactory. The act forces almost all FDA-regulated facilities to implement and validate preventive control plans,[50] which resemble USDA's Hazard Analysis Critical Control Point (HACCP) plans. In addition, FSMA equips FDA with the mandatory recall authority it previously lacked. FSMA contains a number of additional rules important for regulated entities to know and understand, including the Sanitary Transportation Rule,[51] Produce Safety Rule,[52] Foreign Supplier

---

47. *Id.* § 372(a)(1)(A).
48. *Id.* § 374.
49. 35A Am. Jur. 2d *Food* § 50 (2021) (citing United States v. Thriftimart, Inc., 429 F.2d 1006 (9th Cir. 1970); Daley v. Weinberger, 400 F. Supp. 1288 (E.D.N.Y. 1975)).
50. *See* 21 C.F.R. § 117.126 *et seq.* (2021).
51. *See id.* § 1.900 *et seq.*
52. *See id.* § 112.1 *et seq.*

Verification Programs Rule,[53] and Food Defense Rule.[54] FDA has proposed a number of additional rules under FSMA, including a proposed rule on laboratory accreditation[55] and food traceability.[56]

Inspections

Both USDA and FDA primarily seek to keep the general public safe through the use of inspections of regulated entities. Such an inspection constitutes an official examination for a regulatory purpose. Inspections provide USDA and FDA with important information, but the regulators' inspection powers are not limitless. Instead, the agencies must operate within sometimes murky constitutional, statutory, and regulatory boundaries in conducting inspections. The differences between USDA's and FDA's inspections are both dramatic and important. As part of its authority, USDA stations inspectors at food production facilities during all operations to continuously enforce inspection protocols. Those USDA inspectors must be present for any regulated facility to be operating.[57] By contrast, FDA inspectors only periodically visit production facilities under their authority. Although USDA inspectors are present daily in the facilities they regulate, it is common for an FDA-regulated facility to go for several years without an inspection.

In addition to its inspection activities, FSIS may investigate reports of foodborne illnesses potentially tied to USDA-regulated products.[58] FSIS also maintains a Standing Emergency Management Committee, which USDA can activate to investigate nonroutine incidents involving potential adulteration of FSIS-regulated products.[59] Facilities subject to inspection must allow inspectors access to all parts of their facility at all times,

---

53. *See id.* § 1500 *et seq.*
54. *See id.* § 121.1 *et seq.*
55. *See* FDA, *FSMA Proposed Rule on Laboratory Accreditation*, https://www.fda.gov/food/food-safety-modernization-act-fsma/fsma-proposed-rule-laboratory-accreditation (last updated Apr. 3, 2020).
56. *See* FDA, *FSMA Proposed Rule for Food Traceability*, https://www.fda.gov/food/food-safety-modernization-act-fsma/fsma-proposed-rule-food-traceability (last updated Jan. 12, 2021).
57. 9 C.F.R. § 300.2. (2021).
58. Procedures for FSIS foodborne illness investigations are set forth in FSIS Directive 8080.3 (Oct. 27, 2017), https://www.fsis.usda.gov/sites/default/files/media_file/2020-07/8080.3.pdf.
59. Procedures for incident investigation teams are set forth in FSIS Directive 5500.3 (July 19, 2006), https://www.fsis.usda.gov/sites/default/files/media_file/2020-07/5500.3.pdf.

around the clock, and without reference to whether the establishment is operational or not.[60] In addition, all livestock and all products entering any "official establishment" must be inspected. Unlike its FDA counterpart, FSIS inspections are generally performed on a continuous basis with inspectors permanently located in regulated facilities every day. Accordingly, while USDA does not enjoy "mandatory recall" authority like FDA, it does have the ability to unilaterally stop operations at any facility it suspects of violating federal law by pulling inspectors from the facility and refusing to continue inspections.

USDA uses three main enforcement actions against regulated facilities. A "regulatory control action" is when USDA requires retention of product, rejection of equipment or facilities, slowing or stopping of lines, or refusal to allow the processing of specifically identified products.[61] A "withholding action," by contrast, is the refusal to allow inspection marks to be applied to a facility's product, thereby preventing that particular product from ultimate sale to consumers.[62] A "suspension" is USDA's halt of inspection activity, which has the practical effect of shutting down an entire plant.[63]

In stark contrast to USDA's inspectors, FDA inspectors do not set up permanent shop in the facilities they regulate. In practice, FDA may not visit major regulated facilities for several years at a time, to say nothing of smaller entities. Thus, the effect of FDA's inspection differs from USDA's substantially. FDA may periodically enter, at reasonable times, any factory, warehouse, or establishment in which food is manufactured, processed, packed, or held, for introduction into interstate commerce. The agency also has the authority to enter any vehicle used to transport or hold such food.[64] Once inside, FDA inspectors have the right to inspect the entire factory, warehouse, establishment, or vehicle, including all equipment, finished and unfinished materials, containers, and labeling. FDA inspectors are not explicitly authorized to interview establishment employees, but they are also not prevented from such interviews. Inspectors generally provide any interviewed employee with Miranda-style warnings.[65] But, as a practical matter, FDA inspectors can and will engage employees in conversation, and may even ask for written statements.

---

60. 9 C.F.R. § 300.6 (2021).
61. *Id.* § 500.1(a).
62. *Id.* § 500.1(b).
63. *Id.* § 500.1(c).
64. FDA, INVESTIGATIONS OPERATIONS MANUAL 2021 [hereinafter IOM], *available at* https://www.fda.gov/media/113432/download (last visited May 4, 2021).
65. *See* United States v. Gel Spice Co., 601 F. Supp. 1214 (E.D.N.Y. 1985).

Until recently, FDA inspectors did not have the right to inspect financial data, personnel data (other than qualifications of certain personnel), or research data.[66] However, FSMA effectively provided FDA inspectors with the right to review any documents that they deem necessary to prevent the consumption of food reasonably likely to cause serious adverse health consequences, or death to humans or animals.[67] While such documents may be stored off-site or electronically, they must be produced within 24 hours by the regulated entity. Finally, FDA does not normally provide advance notice of an inspection[68] and any inspection must occur at a reasonable time, within reasonable limits, and in a reasonable manner.[69] Preventing or delaying an FDA inspector from entering a facility may be deemed a refusal, which is itself a "prohibited act" under FDCA,[70] and can subject the facility to penalties, including potential criminal sanctions.[71]

Prohibited Acts

After determining which regulatory scheme applies to a company's products and, thus, which regulatory agency oversees it, the entity must next determine what the applicable act prohibits. While each of the regulatory schemes overlaps substantially in what it prohibits, each uses its own verbiage in doing so. And, importantly, entities should be aware of the potential penalties for violating the act (or acts) applicable to their products.

FDCA contains a laundry list of prohibited conduct under the act.[72] Most importantly, however, the act prohibits "the adulteration or misbranding of any food, drug, device, tobacco product, or cosmetic in interstate commerce,"[73] as well as the subsequent "introduction or delivery for introduction into interstate commerce of any drug, device, tobacco product, or cosmetic that is adulterated or misbranded."[74] Both "adulterated" and "misbranded" are terms of art under FDCA and defined by the act. According to FDCA, a food can be adulterated in a number of different

---

66. 21 U.S.C. § 374(a)(1).
67. *Id.* § 350c(a).
68. *See* IOM, *supra* note 64.
69. *Id.*
70. *See* 21 U.S.C. § 331.
71. United States v. Chung's Prods. LLP, 941 F. Supp. 2d 770 (S.D. Tex. 2013) (noting delay of four hours).
72. *See* 21 U.S.C. § 331.
73. *Id.* § 331(b).
74. *Id.* § 331(a).

ways,[75] but the most common usages of the term involves "poisonous, insanitary, etc., ingredients," which "may render it injurious to health."[76] Similarly, FDCA defines "misbranded food" in a number of important ways,[77] including food having a false or misleading label;[78] offered for sale under another name;[79] imitating another food (unless disclaimed per the regulation);[80] sold in a misleading container;[81] and where the information required on the label is not prominently placed on the label so as "to render it likely to be read and understood by the ordinary individual under customary conditions of purchase and use."[82]

If an entity violates FDCA, it may be subject to a number of potential penalties. Most immediately, the act allows FDA to seek an injunction in federal court to restrain the alleged violation or violations.[83] At its most extreme, FDCA calls for a year's imprisonment, a $1,000 fine, or both, for a violation of section 331,[84] with subsequent violations (or violations with an intent to defraud or mislead) being subject to felony charges that can result in three years in prison and a fine of $10,000.[85] While prosecutions of individual employees in the industry were once rare, the past decade has seen a marked increase in such prosecutions.[86] Companies (and their individual employees) can minimize their exposure to these sanctions by obtaining FDCA "guarantees" from the entities supplying them products and ingredients.[87] The act also provides an exception for misbranded food when a violation exists solely because of a product's advertising.[88] Under the act, FDA may seize adulterated or misbranded

---

75. *Id.* § 342.
76. *Id.* § 342(a).
77. *Id.* § 343.
78. *Id.* § 343(a).
79. *Id.* § 343(b).
80. *Id.* § 343(c).
81. *Id.* § 343(d).
82. *Id.* § 343(f).
83. *Id.* § 332(a). The act further defines a violation of any injunction as a further violation of FDCA and subjects the entity to trial by jury. *See id.* § 332(b).
84. *Id.* § 333(a)(1).
85. *Id.* § 333(a)(2).
86. *See* Daniel C. Zinman & Alex M. Solomon, *From Bonds to Burritos: How the Increasing Criminalization of Regulatory Offenses Affects the Food and Beverage Industry*, FOOD SAFETY MAG., May 3, 2016, https://www.food-safety.com/articles/4740-from-bonds-to-burritos-how-the-increasing-criminalization-of-regulatory-offenses-affects-the-food-and-beverage-industry.
87. 21 U.S.C. § 7.14.
88. *Id.* § 333(d).

products,[89] with costs potentially paid by the entity.[90] The act prescribes a number of other penalties, including debarment of entities importing food into the United States,[91] as well as additional civil penalties.[92]

By comparison, FMIA contains a much smaller list of prohibited acts than FDCA.[93] First, entities regulated by the act are prohibited from slaughtering animals for use as human food in any way out of compliance with FMIA's requirements.[94] Second, entities are prohibited from slaughtering animals inhumanely and out of compliance with the Humane Methods of Livestock Slaughter Act.[95] Regulated entities are further prohibited from selling, transporting, or receiving adulterated or misbranded articles or articles requiring inspection but lacking a passing inspection.[96] Finally, the act prohibits any act that results in human food being sold as adulterated or misbranded.[97] Under FMIA, any individual or entity violating the act shall be subject to a year in prison, a $1,000 fine, or both, while violations with an intent to defraud may result in three years in prison and a $10,000 fine.[98]

Similarly, PPIA prohibits certain acts similar to FMIA.[99] For instance, PPIA prohibits slaughtering of poultry in violation of the act,[100] as well as sale and transportation of poultry products capable for use as human food that are either adulterated or misbranded or lacking required inspection.[101] In addition, the act prohibits acts that would result in poultry being sold as adulterated or misbranded.[102] PPIA also prohibits the sale or transportation of slaughtered poultry "from which the blood, feathers, feet, head, or viscera have not been removed" in accordance with USDA regulation.[103] The act further prohibits the use of trade secrets learned under PPIA's auspices for personal gain.[104] Finally, PPIA prohibits the

---

89. *Id.* § 334(a)(1).
90. *Id.* § 334(e).
91. *Id.* § 335a(b)(1)(C).
92. *Id.* § 335b.
93. *See id.* § 610.
94. *Id.* § 610(a).
95. *Id.* § 610(b).
96. *Id.* § 610(c).
97. *Id.* § 610(d).
98. *Id.* § 676(a).
99. *Id.* § 458.
100. *Id.* § 458(a)(1).
101. *Id.* § 458(a)(2).
102. *Id.* § 458(a)(3).
103. *Id.* § 458(a)(4).
104. *Id.* § 458(a)(5).

misuse of USDA inspection marks.[105] Like the other regulatory schemes, PPIA prescribes punishments up to a year in prison and a $1,000 fine for violations, as well as three years imprisonment and a $10,000 fine for attempts to defraud under the act.[106]

Last, under EPIA, regulated entities are prohibited from selling or transporting eggs except as authorized by the act.[107] Further, egg handlers are prohibited from using restricted eggs in the preparation of human food except as prescribed by USDA regulations.[108] EPIA prohibits the processing of eggs outside of compliance with the act,[109] as well as the sale or transportation of eggs not labeled and packaged as required by the act.[110] The act additionally prohibits plants not complying with the act or USDA regulations,[111] and prohibits plants from allowing eggs to be moved from a plant if they are adulterated or misbranded and capable of use as human food.[112] EPIA also mandates that egg handlers store and transport eggs under refrigeration no greater than 45 degrees.[113] The act additionally prohibits various misuses of USDA marks and prohibits the use of trade secrets under the act's ambit for personal gain. EPIA punishes violations with imprisonment up to a year and a fine of $5,000, with violations involving an intent to defraud being subject to a three-year prison sentence and a fine of $10,000.[114] The act further identifies a series of civil penalties for violations in addition to potential criminal liability.[115]

Allergens

Another area of particular concern for food safety specialists is how companies handle potential allergens, including required labeling for foods potentially containing the eight major allergens. A food allergy is a reaction of the immune system triggered by eating certain foods.[116] In the

---

105. *Id.* § 458(b)–(c).
106. *Id.* § 461(a).
107. *Id.* § 1037(a)(1).
108. *Id.* § 1037(a)(2).
109. *Id.* § 1037(b)(1).
110. *Id.* § 1037(b)(2).
111. *Id.* § 1037(b)(3).
112. *Id.* § 1037(b)(4).
113. *Id.* § 1037.
114. *Id.* § 1041(a).
115. *Id.* § 1041(c).
116. *See* Mayo Clinic, *Food Allergy*, https://www.mayoclinic.org/diseases-conditions/food-allergy/symptoms-causes/syc-20355095 (last visited May 4, 2021).

most severe cases, food allergies can trigger *anaphylaxis*, which results in constriction and tightening of airways, swollen throat, shock, and rapid pulse. These severe reactions may require emergency treatment and can result in coma or even death.[117] Approximately one in three children and roughly one in ten adults suffer from food allergies.[118]

A majority of food allergies are caused by proteins in foods, namely proteins in milk, eggs, fish, crustacean shellfish, tree nuts, peanuts, wheat, and soybeans.[119] In an effort to protect Americans suffering from food allergies, Congress passed the Food Allergen Labeling and Consumer Protection Act of 2004,[120] which applies to all FDA-regulated foods. The act requires that food labels must clearly identify the product's ingredients, but also the source of all the ingredients derived from the eight most common food allergens, which account for nearly 90 percent of allergic reactions to food.[121] Those who are allergic to other substances, however, may find their failure-to-warn claims preempted by this same law.[122]

Preventive Controls and Countermeasures

Regulated entities cannot simply rely on agency inspections to ensure that their products remain safe and sanitary for the general public. Indeed, the risk of pathogens to food safety is pervasive. Instead, producers must employ various countermeasures to reduce negative outcomes or mitigate risks that can cause foodborne illnesses. When well designed and implemented, countermeasures reduce the number of food safety illnesses by reducing the potential that a pathogen is present in a product.

Both USDA and FDA require their regulated entities to develop, maintain, and validate proactive food safety programs. For USDA-regulated entities, these systems are referred to as HACCP plans. By

---

117. *Id.*
118. *See* Food Allergy Research & Education, *Facts and Statistics*, https://www.foodallergy.org/resources/facts-and-statistics (last visited May 4, 2021).
119. *See* Mayo Clinic, *supra* note 116.
120. *See* Pub. L. No. 108-282, 118 Stat. 905 (2004).
121. *See* FDA, *Food Allergies*, https://www.fda.gov/food/food-labeling-nutrition/food-allergies (last updated Apr. 26, 2021). *See, e.g., In re* McDonald's French Fries Litig., 503 F. Supp. 2d 953 (N.D. Ill. 2007) (plaintiffs who chose to not consume gluten, milk, and wheat, for allergen or dietary reasons, allege that McDonald's made fraudulent representations that its french fries were free of gluten, milk, and wheat).
122. *See, e.g.*, Cardinale v. Quorn Foods, Inc., No. X05CV096002022S, 2011 Conn. Super. LEXIS 1262 (May 19, 2011) (state court claim alleging failure to warn of mycoprotein (to which some are allergic) was preempted by federal law, which requires no warning against this allergen).

contrast, the vast majority of FDA-regulated entities, at least following the passage of FSMA, are required to develop analogous food safety plans that include "preventive controls." Whether referred to as HACCP or a food safety plan, both constitute safety control systems that emphasize systematic mitigation of risk at points in the production process where those risks are most likely. The emphasis on these systems, rather than sole reliance on inspection of finished products for pathogens or deleterious substances, is relatively new to the food safety world.[123] FSIS began requiring federally inspected meat and poultry plants to implement an HACCP system beginning in 1996.[124] Similarly, while FDA traditionally required producers who make juice or juice concentrates for subsequent beverage use[125] and seafood processors[126] to apply HACCP principles, as of 2015 under FSMA,[127] most FDA-regulated producers will be required to develop and validate food safety plans with preventive controls. The two analogous systems functionally overlap in both intention and effect, but use their own vocabulary.

For instance, according to FSIS, the seven principles underlying a successful HACCP program[128] are as follows:

1. **Conduct a hazard analysis.**[129] Because a hazard cannot be mitigated until it is identified, producers must determine the relevant hazards inherent with any product, facility, and personnel, and also identify preventative measures that can be applied to eliminate or mitigate those hazards. Hazards can include biological agents, chemical contaminants, and physical impurities.

---

123. FDA defines HACCP as "a management system in which food safety is addressed through the analysis and control of biological, chemical, and physical hazards from raw material production, procurement and handling, to manufacturing, distribution and consumption of the finished product." FDA, *Hazard Analysis Critical Control Point (HACCP)*, https://www.fda.gov/food/guidance-regulation-food-and-dietary-supplements/hazard-analysis-critical-control-point-haccp (last updated Jan. 29, 2018).
124. 9 C.F.R. §§ 304, 308, 310, 327, 381, 416, 417 (2021).
125. 21 C.F.R. § 120 *et seq.* (2021).
126. *Id.* §§ 123, 1240; United States v. N.Y. Fish, Inc., No. 13-cv-2909, 2014 U.S. Dist. LEXIS 42716 (E.D.N.Y. Mar. 30, 2014); United States v. Blue Ribbon Smoked Fish, Inc., 179 F. Supp. 2d 30 (E.D.N.Y. 2001).
127. 21 U.S.C. § 2201 *et seq.*
128. University of Nebraska-Lincoln Institute of Agriculture and Natural Resources, *The Seven Principles of HACCP*, https://food.unl.edu/seven-principles-haccp (last visited May 27, 2021).
129. *Id.*

2. **Identify the critical control points (CCPs).**[130] CCPs are a step or procedure in a production or preparation process where a hazard can be prevented, eliminated, or reduced to acceptable levels, such as pasteurization.
3. **Establish critical limits for each CCP.**[131] A critical limit is the minimum or maximum value at which the presence of a hazard is acceptable, such as the maximum temperature at which a pathogen can survive. Companies should seek to create critical limits that are quantifiable and easily verifiable, such as a time/temperature requirement that an ingredient be subjected to a specific temperature for a specific time rather than vesting a facility employee with the discretion to determine whether the ingredient appears appropriately cooked.
4. **Establish monitoring procedures.**[132] Each CCP must be monitored by the company to ensure its mitigation measures are effective in reducing hazards to their critical limits. When possible, entities should employ automated alarms or stops that automatically cease production when critical limits are not achieved.
5. **Establish corrective actions.**[133] Companies should identify actions to be taken if monitoring indicates that a CCP has failed to prevent a deviation from a critical limit, or if, despite achieving the critical limit, a hazard still exists. Corrective actions are fail-safes designed to prevent dangerous products from reaching consumers following a control failure.
6. **Establish recordkeeping procedures.**[134] Producers must maintain certain documents, including hazard analyses and written HACCP plans, monitoring records for CCPs, determinations regarding critical limits, records of verification activities, and records of corrective actions. These records often provide crucial evidence in responding to regulatory inquiries and in rebutting litigation claims.
7. **Establish verification procedures.**[135] When an HACCP plan becomes ineffective, continued maintenance and adherence to the plan does little good for a company. To ensure continued

---

130. *Id.*
131. FSIS, USDA, GUIDEBOOK FOR THE PREPARATION OF HACCP PLANS 11 (2020).
132. *Id.* at 12.
133. *Id.* at 13.
134. *Id.* at 14.
135. *Id.* at 15.

efficacy of an HACCP plan, companies should engage in a two-step process:

- **Validation** is a first step wherein the HACCP system is tested and reviewed by the company. Testing offers evidence that the plan adopted by the producer works effectively under actual production conditions. Validation must be performed by someone with HACCP training. For instance, validation may include a test pathogen being intentionally added to a designated product that then goes through the CCP to determine whether the procedure consistently removes the spoilage.[136]
- **Verification**, by contrast, is the ongoing assessment by a producer used to ensure that the HACCP plan is working as designed on a daily basis. Verification steps includes monitoring instrument calibration, observing activities designed to monitor efficacy, and reviewing required records for compliance.

By contrast, FDA mandates that entities covered by FDCA develop and implement written food safety plans.[137] Consistent with FDA regulations, a compliant food safety plan includes the following components:

1. **Hazard analysis**.[138] First, entities must determine which hazards potentially threaten the food product or products at a facility. These "known or reasonably foreseeable" hazards could be biological, chemical, or physical, and could occur naturally, be unintentionally introduced into the product, or even intentionally introduced for economic gain. If a company's analysis reveals one or more potential hazards, the producer must then develop and implement preventative controls for each hazard.
2. **Preventive controls**.[139] While producers have flexibility in developing specific preventive controls, ultimately, a facility must have written preventive controls in place for the hazards identified for each product manufactured at the facility. Preventive

---

136. *See* Paramount Farms, Inc. v. Ventilex, B.V., No. CV f 08-1027 LJO SKO, 2011 U.S. Dist. LEXIS 1902 (C.D. Cal. Jan. 3, 2011) (describing failed validation efforts in thermal processing unit designed to achieve "five-log" reduction of pathogens in almonds and pistachios).

137. *See* 21 C.F.R. § 117.126 *et seq.* (2021).

138. *See* FDA, *FSMA Final Rule for Preventive Controls for Human Food*, https://www.fda.gov/food/food-safety-modernization-act-fsma/fsma-final-rule-preventive-controls-human-food (last updated July 10, 2020).

139. *See id.*

controls must ensure that identified hazards are significantly minimized or prevented entirely. Further, preventive controls must help ensure that products are not adulterated. FDA regulations identify the following preventive controls:

- **Process controls** are controls that ensure control parameters are met. Examples of process controls include cooking, refrigerating, and acidifying. These controls must include parameters and values like HACCP's critical limits, which are appropriate for the control and its role in the facility's plan.
- **Food allergen controls** constitute those written procedures implemented by a facility both to prevent allergen cross-contamination and to ensure allergen ingredients are appropriately disclosed on packaged food labels.
- **Sanitation controls** are the procedures, practices, and processes implemented by a facility to ensure it remains in a sanitary condition and to prevent or minimize sanitation hazards. Such hazards could include environmental pathogens, hazards resulting from employees handling food, and allergens.
- **Other controls** are those controls not enumerated elsewhere, but that are deemed necessary by a producer to ensure a hazard at a facility is significantly minimized or prevented.

3. **Oversight and management.**[140] After identifying preventive controls for each facility hazard, companies must next implement procedures to ensure the control measures are being used and are effective. Those procedures can include:

- **Monitoring** procedures are those that a facility uses to ensure preventive controls are being performed consistently and may differ depending on the preventive control being monitored. As an example, a facility may monitor a heating process to kill potential pathogens by recording temperature values at regular intervals. Monitoring procedures should be documented and retained.
- **Corrections** are the procedures undertaken by a facility to both identify and correct minor issues that occur during production. Facilities should use corrections for nonsystemic problems that arise.

---

140. *Id.*

- **Corrective actions** constitute actions taken to identify and resolve problems in implementing a preventive control. A corrective action should reduce the likelihood of a recurrence of a problem; assess the affected food(s) for safety issues; and prevent adulterated food from entering the stream of commerce. Facilities must document and retain corrective actions taken.
- **Verifications** are required procedures to ensure that the facility's preventive controls are being implemented consistently and remain effective against identified hazards. For instance, a facility could scientifically validate its preventive controls to ensure they remain capable of controlling for a hazard. Companies could also calibrate and verify various instruments used in process controls to confirm that they are providing accurate and reliable readings. In cases where a facility processes ready-to-eat food, that facility must engage in environmental monitoring when it has identified environmental pathogens as a potential hazard.

4. **Supply chain**.[141] Companies should create and implement risk-based supply chain programs in the event that they identify a hazard that (1) requires a preventive control, and (2) the company will apply the control within the facility's supply chain. However, a company need not develop supply chain controls if either they control the hazard within their own facility or if a subsequent entity within the supply chain will control the hazard. In addition, FDA requires that manufacturers receive raw materials and other ingredients requiring supply chain controls from approved suppliers, or only on a temporary basis from unapproved suppliers provided that the materials are verified before use. To the extent a company relies on another entity within the supply chain for supplier verification (e.g., a broker or distributor), the facility must still review that entity's control documentation.
5. **Recall plan**.[142] Finally, if a facility determines that an identified hazard requires a preventive control, the facility must also develop and maintain a written recall plan. In the plan, the company must describe the procedures for performing a recall of the product, as well as procedures to notify consignees and the public (as necessary), to conduct effectiveness checks, and procedures to dispose of recalled product.

---

141. *Id.*
142. *Id.*

Chapter 5

Food Recalls

James F. Neale and Benjamin P. Abel

This chapter will explore the legal framework governing food recalls, recommend practices for conducting a successful recall, and discuss post-recall considerations.

Governmental Agencies Responsible for Food Product Recalls

At the federal level, two agencies primarily regulate food safety, the U.S. Food and Drug Administration (FDA) and the U.S. Department of Agriculture (USDA). As a general matter, USDA regulates the safety of meat, poultry, and egg products, while FDA is responsible for the safety of all other food products.

The Centers for Disease Control and Prevention (CDC) monitors public health and alerts other regulatory agencies in the event of a suspected foodborne illness outbreak via the Foodborne Diseases Active Surveillance Network (FoodNet), the National Antimicrobial Resistance Monitoring System for Enteric Bacteria (NARMS), and the National Molecular Subtyping Network for Foodborne Disease Surveillance (PulseNet).[1] Taken together, FoodNet, NARMS, and PulseNet provide

---

1. For information on CDC, FoodNet, PulseNet, and NARMS, see https://www.cdc.gov/, https://www.cdc.gov/foodnet/, https://www.cdc.gov/pulsenet, and https://vww.cdc.gov/narms/, respectively.

information to FDA and USDA to identify the sources of illnesses and, as necessary, to recall the affected product.

Regulators Governing Food Product Recalls

Most people are familiar with the concept of food recalls. However, exactly how food product recalls are initiated is often misunderstood. Stories in the news often state that FDA or USDA "ordered" a particular recall. However, until 2011's enactment of the Food Safety Modernization Act,[2] neither FDA nor USDA had the authority to "order" a food product recall, excepting the limited ability to order recalls of adulterated or misbranded infant formula.[3] Title 21, part 7, of the *Code of Federal Regulations* lays out FDA's enforcement policy for removing or correcting food products that are distributed despite being violative of laws and regulations administered by the agency. That regulation outlines recall procedures remarkably similar to those implemented by USDA's Food Safety and Inspection Service (FSIS).[4] At least in name, though, food recalls are almost always voluntarily initiated by a food's producer or distributor. Practically, though, the difference between the agencies' abilities to "encourage" a recall rather than mandate one is insignificant. Ultimately, whenever an agency decides a recall is needed, it can force one to occur, even if technically "voluntarily."

Usually, a company discovers a problem and, as a result, decides to initiate a recall of a product on its own. Far less often, a company recalls a product after a formal request for a recall, which, per FDA and FSIS policy, is a step only "reserved for urgent situations."[5] Regardless of how the agency initiates a recall, the agencies' role is ultimately to oversee the company's strategy and the efficacy of the recall. Since FDA's and FSIS's recall policies are similar, and because FDA is overwhelmingly responsible for the U.S. food supply, this chapter will focus primarily on FDA recalls, except where FSIS's treatment is meaningfully different.[6]

---

2. 21 U.S.C. § 2201 *et seq.*
3. Infant Formula Act of 1980, codified at 21 U.S.C. § 321.
4. *See generally* DOUG ARCHER ET AL., UNIVERSITY OF FLORIDA, THE FOOD RECALL MANUAL 19 (2018) [hereinafter THE FOOD RECALL MANUAL] (noting that "[21 C.F.R. part 7] outlines the recall procedures to be used either by the FDA or FSIS").
5. 21 C.F.R. § 7.40 (2021).
6. FSIS food recall policies can be found in FSIS Directive 8080.1 v.5 (Nov. 17, 2008) [hereinafter FSIS Directive 8080.1]. Although FSIS's recall policies largely

Recall Types

Company-Initiated Recalls

Often, a company will decide to initiate a recall of a product from the marketplace based on information learned from a nongovernmental source. As an example, a manufacturer may find out that a machine on its production line discharged metal flakes into the product. Or, a company might receive customer or retailer complaints that its products taste or appear "off." Per FDA guidelines, a manufacturer may decide on its own to remove a distributed product in these circumstances.[7] However, just because a product is removed from the marketplace does not necessarily mean a "recall" has occurred.[8] Instead, any company believing the product(s) it is removing from distribution violates federal food laws "is requested" to notify the appropriate FDA district office immediately.[9] Such a removal or correction will be considered a "recall" only if it addresses a violation that is subject to legal action (e.g., seizure).[10]

Other times, FDA might inform a manufacturer that it has determined a product violates or may violate the law.[11] Recalls initiated after such information is received are still treated as voluntary "firm-initiated" recalls even if the company feels like it has little choice but to initiate the process.

In an instance where FDA deems a product's removal from the market a recall, the firm will be asked to provide it with specific information, such as (1) the identity of the product involved;[12] (2) the reason for the removal and the date and circumstances in which the product deficiency or possible deficiency was discovered;[13] (3) an evaluation of the risk associated with the product;[14] (4) the total amount of products produced or the time span of the production, or both;[15] (5) the total amount of such

---

parallel those promulgated by FDA in 21 C.F.R. § 7 *et seq.*, there are some substantive differences.

7. 21 C.F.R. § 7.46(a) (2021).
8. *Id.*
9. *Id.* ("A firm that [removes product from the market] because it believes the product to be violative is requested to notify immediately the appropriate Food and Drug Administration district office listed in [21 C.F.R.] § 5.115").
10. *Id.*
11. *Id.*
12. *Id.* § 7.46(a)(1).
13. *Id.* § 7.46(a)(2).
14. *Id.* § 7.46(a)(3).
15. *Id.* § 7.46(a)(4).

products estimated to be in distribution channels;[16] (6) the number and identity of direct accounts;[17] (7) a copy of the firm's recall communication if any has issued, or a proposed communication if none has issued;[18] and (8) a proposed strategy for conducting the recall.[19] With this information in hand, FDA will then advise the manufacturer of the recall's classification[20] and the agency's recommended changes, when appropriate, to the company's strategy for the recall.[21] A firm facing a potential recall should not wait for FDA to complete its review before taking steps to remove the product from the market.[22]

Government-Initiated Recalls

In "urgent situations," FDA may decide to formally request a recall. FDA will only initiate such a recall when (1) a product that has been distributed presents a risk of illness or injury or gross consumer deception; (2) the firm has not initiated a recall of the product (usually after a "request" from FDA); and (3) an agency action is necessary to protect the public health.[23] In such instances, FDA will notify the firm in writing that the agency has determined a recall should be initiated immediately.[24] FDA's notification will specify the violation, the health hazard classification of the subject product, the recall strategy, and other necessary instructions.[25] When the agency issues a formal recall request, the manufacturer often has fewer opportunities to provide input regarding the recall strategy than it would had it engaged in a firm-initiated recall.

Manufacturer's Failure or Refusal to Initiate Voluntary Recall

If a company refuses to recall a product, FDA has several options, including (1) bringing a civil action enjoining the manufacturing process; (2) seizing existing products; or (3) bringing a criminal action for violation of certain Federal Food, Drug, and Cosmetic Act (FDCA) provisions.[26] The FDCA provides specific authorization for district courts to

---

16. *Id.* § 7.46(a)(5).
17. *Id.* § 7.46(a)(6).
18. *Id.* § 7.46(a)(7).
19. *Id.* § 7.46(a)(8).
20. *See infra* Recall Classifications section.
21. 21 C.F.R. § 7.46(b) (2021).
22. *Id.*
23. *Id.* § 7.46(a).
24. *Id.* § 7.46(b).
25. *Id.*
26. 21 U.S.C. § 301 *et seq.*

grant injunctive relief to enforce the act's provisions,[27] and FDA may use that authority to enjoin, both temporarily and permanently, the further production of food violating the FDCA.[28] Practically speaking, a well-thought-out recall will almost certainly make these actions unnecessary.

The agency can additionally decide to use nonjudicial measures, including issuing warning letters and press releases, or including the company in its weekly FDA enforcement report. The effect of such an action on a manufacturer's sales does not constitute a regulatory taking.[29] USDA has similar powers available under the Federal Meat Inspection Act of 1906,[30] Poultry Products Inspection Act of 1957,[31] and Egg Products Inspection Act.[32] As one example, rules promulgated under the Federal Meat Inspection Act allow the agency to detain products violating the act for up to 20 days, to seek a court injunction seizing and condemning such products, or both.[33] In addition to FDA's powers, FSIS has permanent on-site agents at beef and poultry processing facilities to conduct enforcement activities. Those on-site agents have additional enforcement tools, including the ability to issue warnings and citations, and the ability to refuse to grant the mark of inspection (the absence of which is normally crippling to a producer and is the practical equivalent of a "mandatory recall"). These on-site inspectors are responsible for monitoring a plant's compliance with the company food safety plans.[34] Products subject to inspection under the act may neither be sold nor offered for sale in commerce unless they have been both inspected and deemed satisfactory by inspectors.[35] Consequently, when FSIS decides it will no longer

---

27. *Id.* § 332(a).
28. *See, e.g.*, United States v. Union Cheese Co., 902 F. Supp. 778 (N.D. Ohio 1995) (granting preliminary injunction against cheese factory because of presence of *L. monocytogenes*); *see also* United States v. Syntrax Innovations, Inc., 149 F. Supp. 2d 880 (E.D. Mo. 2001) (granting government's motion on summary judgment for misbranding in violation of the FDCA and imposing permanent injunction barring plant operation until specific steps are taken to remedy condition).
29. *See* Dimare Fresh, Inc. v. United States, 808 F.3d 1301 (Fed. Cir. 2015).
30. 21 U.S.C. § 601 *et seq.*
31. *Id.* § 451 *et seq.*
32. *Id.* § 1031.
33. *See, e.g.*, 9 C.F.R. § 329.1 (2021) describing food products and livestock subject to temporary administrative detention); *id.* § 329.6 (stating that any meat or meat food product transported in commerce is subject to seizure and condemnation in a judicial proceeding if such article or livestock is capable of use as human food and is adulterated or misbranded or otherwise is in violation of the act).
34. *Id.* §§ 416.17, 417.8.
35. 21 U.S.C. § 610(c).

conduct inspections and withdraws from a manufacturer, thereby refusing to inspect and pass the company's products, those products may not be sold.

Recall Classifications

FDA (not the company initiating a recall) will assign a "recall classification" or numerical designation—that is I, II, or III—"to indicate the relative degree of health hazard presented by the product being recalled."[36] The agency will evaluate the potential health hazard and will take into account, among other factors, (1) whether any disease or injuries have already occurred from the use of the product;[37] (2) whether the food product could expose humans or animals to a health hazard;[38] (3) an assessment of the hazard to various segments of the population, for example children, the elderly, and so on, who are expected to be exposed to the product being considered, with a focus on those at greatest risk;[39] (4) an assessment of the degree of seriousness of the health hazard;[40] (5) an assessment of the likelihood of occurrence of the hazard;[41] and (6) an assessment of the consequences (immediate or long-range) of occurrence of the hazard.[42]

Class I Recalls

FDA will designate a Class I recall in "a situation in which there is reasonable probability that the use of, or exposure to, a violative product will cause serious adverse health consequences or death."[43] Since most pathogens or allergens can cause serious health consequences to some, certain contaminations automatically warrant a Class I recall. By example, occurrences of Listeria monocytogenes (or other human pathogen) contamination in ready-to-eat foods, or undeclared allergens that are common causes of serious allergic reactions (such as undisclosed shellfish, tree nuts, or soy) normally result in a Class I recall.[44]

---

36. 21 C.F.R. § 7.3 (2021).
37. *Id.* § 7.41(a)(1).
38. *Id.* § 7.41(a)(2).
39. *Id.* § 7.41(a)(3).
40. *Id.* § 7.41(a)(4).
41. *Id.* § 7.41(a)(5).
42. *Id.* § 7.41(a)(6).
43. *Id.* § 7.3.
44. *See generally* THE FOOD RECALL MANUAL, *supra* note 4, at 20 (listing examples of product recalls that will traditionally give rise to a Class I recall).

Class II Recalls

Regulations define a Class II recall as occurring when there "is a situation in which use of, or exposure to, a violative product may cause temporary or medically reversible adverse health consequences or where the probability of serious adverse health consequences is remote."[45] A Class II recall might include a food that requires refrigeration but is not so labeled or an undeclared allergen (not among the "Big 8" allergens).[46] Likewise, beef that potentially carried mad cow disease was designated a Class II recall because there was merely a "remote possibility" that the product would cause illness in humans.[47] The mad cow recall constituted an instance of FSIS considering a Class II recall appropriate even when the potential foodborne illness was serious but the "probability [of negative health consequences] . . . remote."[48]

Class III Recalls

Last, a Class III recall "is a situation in which use of, or exposure to, a violative product is not likely to cause adverse health consequences."[49] Class III recalls might involve nonorganic foods mistakenly labeled as organic or a diet soft drink erroneously labeled as regular.[50] Put simply, Class III recalls involve products that may not technically comply with applicable requirements but carry a minute risk of harm to a consumer.

Product Withdrawals

In some cases involving minor food quality issues, a manufacturer may decide to undertake a market withdrawal, not a recall. Unlike a recall, market withdrawals address "a firm's removal or correction of a distributed product which involves a minor violation that would not be subject to legal action by FDA or which involves no violation."[51] A withdrawal often will involve a product that does not meet the manufacturer's standards or specifications.

---

45. 21 C.F.R. § 73 (2021).
46. *See generally* THE FOOD RECALL MANUAL, *supra* note 4, at 107 (listing examples of product recalls that will traditionally give rise to a Class II recall).
47. Jen Pifer, *USDA Orders Recall of 143 Million Pounds of Beef*, CNN, Feb. 18, 2008, http://www.cnn.com/2008/HEALTH/02/17/beef.recall/index.html.
48. *Id.*
49. 21 C.F.R. § 7.3 (2021).
50. *See generally* THE FOOD RECALL MANUAL, *supra* note 4, at 21 (listing examples of product recalls that will traditionally give rise to a Class III recall).
51. 21 C.F.R. § 7.3 (2021).

Recall Plan and Team

The demands faced by a company making a recall decision require the engagement of a number of individuals on a recall team, ideally created before the need to make recall decisions arises. A company's recall team may consist entirely of in-house employees or may include outside experts, consultants, or counsel. A comprehensive recall team likely includes:

1. **Senior operations personnel.** This should be someone familiar with the company's manufacturing process and traceability. Both are critical in evaluating the cause and extent of potential health hazards. This will often be a plant production manager.
2. **Scientific advisors.** These may include microbiologists, epidemiologists, statisticians, or other scientists who are able to evaluate threats from foodborne pathogens or other contaminants, as well as assess the statistical and scientific evidence on which recall decisions are made.
3. **Logistics or distribution manager.** This person should understand how and where the firm's products are shipped, tracked, and distributed. This individual should be able to provide guidance regarding the appropriate scope of a recall.
4. **Quality assurance specialist.** This person ensures that products meet the company's quality requirements, both as set forth by law and by the manufacturer's guidelines. In addition, this specialist will likely have information about product testing, customer complaints, and related data.
5. **A lawyer.** Counsel is necessary for assessing potential product liability issues, negotiating with insurers about policy coverage or suppliers (such as raw product suppliers) that may bear responsibility for either the recall itself or may otherwise owe indemnity, and responding to government investigations and regulatory activities.[52]
6. **An ultimate decision maker.** Because the committee is composed of various stakeholders with potentially competing interests, consensus is often lacking. As a result, having a company's chief executive officer or another designated decision maker present may vest decision-making power in one individual at a time when quick, decisive decisions are at a premium.

---

52. *See generally* THE FOOD RECALL MANUAL, *supra* note 4, at 28 (recommending a recall team consisting of senior operations manager, public relations specialist, marketing specialist, scientific advisor, logistics specialist, quality assurance specialist, accountant, and attorney).

Other potential members may include accountants, sales managers, public relations specialists, marketing and advertising personnel, and the firm's head of security.[53] To the extent possible, each team member should have access to real-time information detailing data that might be relevant to the recall decision. For example, a manufacturer's quality assurance representative should possess testing and consumer feedback information; a production manager should possess maintenance documents relevant to the company's preventive controls or Hazard Analysis and Critical Control Point (HACCP) plan; a human resources representative should have data concerning employee illnesses at the plant; and so forth.

The ultimate aim of the recall team is to work collaboratively to identify potential threats to product quality so as to avoid the need for a recall. Second, the team should work to develop the company's recall plan in the event a recall is actually necessary. This recall plan should be substantially more comprehensive than the specific "recall strategy" a company provides to FDA as a blueprint for the execution of a specific product recall.[54] Instead, a company's internal recall plan should include notification procedures, and (1) a decision tree, specific to the firm and industry, for evaluating reported product issues to determine whether they warrant corrective action;[55] (2) an action log detailing all actions taken by the recall team, including evaluation of potential product issues;[56] (3) a recall contact list, including upstream suppliers, downstream customers, appropriate government agencies, outside consultants, attorneys, and media;[57] (4) a protocol for collecting and securing records necessary for internal and governmental investigations;[58] (5) a protocol for securing potentially affected product, including ceasing distribution, placing holds on warehoused goods, and notifying customers and wholesalers not to distribute

---

53. *Id.*
54. *See* 21 C.F.R. §§ 7.40–7.59 (2021).
55. THE FOOD RECALL MANUAL, *supra* note 4, at 41–45.
56. *Id.*
57. *Id.*
58. *Id.*

the product;[59] and (6) template drafts of various documents for distribution to customers, government agencies, and media.[60]

In every instance in which information comes to the company's attention (whether from a supplier, consumer, or some other source), the resulting recall report should be documented. Cataloguing all this information will allow the team to respond to regulators' inquiries, to assess potential threats posed by a suspected problem or problems, and, finally, to determine whether it warrants a product recall or some lesser corrective action, such as a market withdrawal. By comprehensively documenting all this information, the recall team will be able to create a record in support of both its decision and the timing of its decision.

However, simply having a plan on a shelf collecting dust does a manufacturer little good. Instead, the company should routinely conduct mock recalls replicating its recall plan, with debriefings following afterwards. Following these mock recalls, the team should revise its plan to build upon what works and excise what does not.

Conducting a Recall

As part of any recall, a specific "recall strategy" is prepared and approved by the relevant agency, either FDA or USDA. A recall strategy addresses not only the depth of a recall, but also the need for public warnings and the extent of effectiveness checks for the recall.[61] Although details of the recall strategy necessarily vary depending upon the product at issue and its scale of distribution, a firm should not wait until the need for a recall strategy arises to start planning. A sufficiently detailed recall plan should include a proposed recall strategy, details of which can be adapted to particular circumstances as they arise.

In a firm-initiated recall, the company will draft the proposed strategy and submit it to FDA for the agency's approval; in an agency-initiated recall, FDA will include a recall strategy in its formal recall request.[62] In both cases, the company's recall strategy should be focused to specifi-

---

59. FSIS's Product Recall Guidelines for Firms states that all FSIS-regulated firms should have a coding system that allows for positive product identification and facilitates effective recall. Records should be maintained "for a period of time that exceeds the shelf life and expected use of the product and at least the length of time specified in FSIS regulations concerning record retention" (9 C.F.R. §§ 320, 381.175 (2021)). *See* FSIS Directive 8080.1, *supra* note 6, attachment 1, § e.

60. *See generally* THE FOOD RECALL MANUAL, *supra* note 4, at 44.

61. 21 C.F.R. § 73 (2021).

62. *Id.* § 7.42(a).

cally address the particular circumstances of each recall based on (1) the results of health hazard evaluation (the recall classification);[63] (2) the ease in identifying the product;[64] (3) the degree to which the product's deficiency is obvious to the consumer or user;[65] and (4) the degree to which the product remains unused in the marketplace.[66]

The recall strategy will specify an appropriate "depth level" for that particular recall.[67] Depth levels can be a wholesale recall (merely recalling products at the wholesaler level, but allowing retailers and ultimate consumers to retain the product),[68] a retail recall (requiring recovery of products from retailers and wholesalers, but not from consumers),[69] or a consumer recall (requiring a recall of all affected products).[70] In addition, the strategy should specify whether a "public warning" is appropriate and, if so, the contents and dissemination of any such warning.[71] As an example, and depending on the circumstances, a particular recall strategy could call for no public warning,[72] a public warning issued to specific trade and industry publications,[73] or a press release to the mainstream media intended to reach individual consumers.[74]

---

63. *Id.* § 7.42(a)(1)(i).
64. *Id.* § 7.42(a)(1)(ii).
65. *Id.* § 7.42(a)(1)(iii).
66. *Id.* § 7.42(a)(1)(iv). The regulations also require consideration of the "continued availability of essential products," but this requirement applies more clearly to FDA's role in regulating pharmaceuticals and medical devices, where the absence of a product from the market could, itself, present a public safety risk. *Id.* § 7.42(a)(1)(v).
67. *Id.* § 7.42(b)(1).
68. *Id.* § 7.42(b)(1)(i).
69. *Id.* § 7.42(b)(1)(ii).
70. *Id.* § 7.42(b)(1)(iii). FSIS policy creates a fourth depth level for product recalls from hotels, restaurants, and other institutional purchasers. *See* FSIS Directive 8080.1, *supra* note 6, attachment 1, § f(A) (defining "HRI" level recall depth).
71. 21 C.F.R. § 7.42(b)(2) (2021).
72. *Id.*
73. *Id.* § 7.42(b)(2)(i).
74. *Id.* § 7.42(b)(2)(ii). It should be noted that FSIS policy calls for a public announcement of the recall to media outlets for all Class I and Class II recalls unless circumstances dictate that such a recall is unnecessary—for example, when the product was not distributed beyond the wholesale level and the firm can recover all affected product without a public announcement. Contrarily, FSIS policy states that a Class III recall will not be accompanied by a public announcement unless circumstances so dictate. *See* FSIS Directive 8080.1, *supra* note 6, § IX(A).

The Recall's Scope

A part of the company's recall plan should establish methods for determining which products are potentially affected, and how to define the class of products subject to the recall. An appropriate product recall may be as small as a particular pallet of product sent to a local retailer or as large as the entire plant's production output during a period of several years. The recall team's operations and logistics experts will be key in determining the scope of any recall. However, it should be noted that in the case of meat and poultry recalls overseen by FSIS, "when the problem involves contamination with microbial pathogens, FSIS generally considers all products produced under a single HACCP plan between performance of complete cleaning and sanitizing procedures (clean-up to clean-up) to be potentially involved."[75] As a result, among others, it is essential that all breaks in production and sanitation efforts be carefully and fully documented, that those documents be kept, and that they be made available to the recall team early in its deliberations.

It is almost impossible to have a "perfectly tailored" recall—that is, to recall all affected product and none that is unaffected. Consequently, the appropriate recall scope is often required to be overbroad. No company wants to recall perfectly good, unaffected product, but public health concerns require firms to err on the side of over-inclusiveness. While a broad recall is undoubtedly expensive, it is both more likely to protect public health and easier to defend in following litigation, if necessary.

While it may be appealing to define a recall as narrowly as possible, there are significant risks in doing so. First, if the initial recall is defined too narrowly, an expansion of the recall may be necessary.[76] In that time, additional consumers may be exposed to the adulterated or mislabeled products. Such a scenario may expose a company to significant, additional liability. Aside from the formal litigation exposure, an expanded recall not only extends the period of negative media coverage, it may further undercut consumers' long-term trust in the company and its products.

On the other hand, an unnecessarily overbroad recall may negatively impact the business, as well. As an example, a recall of food products or ingredients likely triggers various indemnification requirements from wholesalers, retailers, and other manufacturers that incorporate the

---

75. *See* FSIS Directive 8080.1, *supra* note 6, attachment 1, § d.
76. *See, e.g.*, Neb. Beef, Ltd. v. Meyer Foods Holdings, LLC, No. 8:09-cv-00043, 2011 U.S. Dist. LEXIS 23284 (D. Neb. Feb. 24, 2011) (noting that plaintiff had to expand its recall of 500,000 pounds of ground beef to more than five million pounds).

recalling company's ingredients into its own products. Such losses are typically not insured under standard commercial liability policies and may result in indemnification obligations.[77]

Recall Communications

A company engaging in a recall must promptly notify each of its affected direct accounts about the recall.[78] The format, content, and extent of a recall communication should be commensurate with the hazard of the product being recalled and the strategy developed for that recall.[79] The purpose of a recall communication is to convey (1) that the product in question is subject to a recall;[80] (2) that further distribution or use of any remaining product should cease immediately;[81] (3) where appropriate, that the direct account should in turn notify its customers who received the product about the recall;[82] and (4) instructions regarding what to do with the product.[83]

A company's recall communication should be immediately disseminated electronically, but can also be done by telegrams, mailgrams, or first-class letters conspicuously marked, preferably in bold, red type, on the letter and the envelope with "food recall."[84] This letter and envelope should also be marked "urgent" for Class I and Class II recalls and, when appropriate, for Class III recalls.[85] Phone calls or other personal contacts should be confirmed by one of the above methods or documented in an appropriate manner, or both.[86] The communication should (1) be brief and to the point;[87] (2) provide accurate and immediate identification of the product;[88] (3) explain concisely the reason for the recall and the hazard involved;[89] (4) provide specific instructions about what should be

---

77. *See, e.g.*, Acme Steak Co., Inc. v. Great Lakes Mech. Co., No. 98-C.A.-146, 2000 Ohio App. LEXIS 4578 (Ohio Ct. App. Sept. 29, 2000).
78. 21 C.F.R. § 7.49(a) (2021).
79. *Id.*
80. *Id.* § 7.49(a)(1).
81. *Id.* § 7.49(a)(2).
82. *Id.* § 7.49(a)(3).
83. *Id.* § 7.49(a)(4).
84. *Id.* § 7.49(b).
85. *Id.*
86. *Id.*
87. *Id.* § 7.49(c)(1)(i).
88. *Id.* § 7.49(c)(1)(ii) (including "product, size, lot number(s), code(s) or serial number(s) and any other pertinent descriptive information to enable accurate and immediate identification of the product").
89. *Id.* § 7.49(c)(1)(iii).

done with the recalled products;[90] and (5) provide means for the recipient to report to the recalling firm whether it has any of the product.[91]

Sophisticated product distribution systems often allow for a centralized, automated lockout that, once activated, prevents the further shipping, distribution, or sale of recalled products. To the greatest possible extent, that system should be in place so that, with the press of a button, recalled product cannot be released from a distribution center, or purchased from a retailer's checkout register. Automating such procedures can also preserve public health and limit the number of potential claims.

Status Reports

After a product recall is underway, the recalling company is required to submit periodic "recall status reports" to FDA.[92] FDA will determine the frequency of those reports in light of the urgency of the recall. However, a reporting interval between two and four weeks is typical.[93] Unless FDA otherwise requires, a status report should include the following information: (1) the number of consignees notified of the recall, and when and how they were notified;[94] (2) the number of respondents to the recall communication and the amount of products held at the time the recall communication was received;[95] (3) the number of consignees that did not respond;[96] (4) the number of products returned or corrected by each consignee contacted and the quantity of products accounted for;[97] (5) the number and results of effectiveness checks performed;[98] and (6) the estimated time frames for completion of the recall.[99] A company's reporting obligations run until the recall is terminated by FDA.[100]

Regulator Site Inspection Contemporaneous with Recall

After being informed of a product recall, FDA's district office will reach out to the firm to obtain necessary information and generally arrange

---

90. *Id.* § 7.49(c)(1)(iv).
91. *Id.* § 7.49(c)(5)(v).
92. *Id.* § 7.53(n).
93. *Id.*
94. *Id.* § 7.53(b)(1).
95. *Id.* § 7.53(b)(2).
96. *Id.* § 7.53(b)(3) (if needed, FDA will request the identity of non-responding consignees).
97. *Id.* § 7.53(b)(4).
98. *Id.* § 7.53(b)(5).
99. *Id.* § 7.53(b)(6).
100. *Id.* § 7.53(c).

to commence an on-site investigation.[101] If the recall is classified as, or appears to be, a Class I or a significant Class II recall, FDA's internal procedures state that an establishment inspection should take place to, among other things, determine the root causes of the problem and document violations for possible regulatory action.[102] That inspection serves a number of different purposes depending on the nature and scope of the recall. As an example, at the beginning, the inspection will focus on identifying the affected product and containing it. This information will prove necessary in determining an appropriate recall scope and strategy given the public health threat. Subsequent steps in the inspection, however, may focus on assuring that the cause of the product recall has been remedied and that there is no longer any danger in manufacturing and releasing products from that facility. Likewise, the length of the site inspection may be short and last a day, or could extend for weeks, months, or longer, depending on the recall's scope, the nature of the threat, and whether the subject facility will return to full operations. In a food product recall, FDA/USDA's inspection should:

- Obtain the firm's proposed recall strategy[103]
- Collect copies of all relevant labeling associated with the affected product[104]
- Obtain a complete distribution list of all shipments of the suspect lots of products[105]
- Obtain supporting documentation to assist the agency in identifying and evaluating the problem or problems, such as product complaints and specifications, and test results, including the methods used to obtain the results[106]
- Assess the root cause or causes of the problem, including determining how and when the problem occurred and how and when it was discovered, and obtaining the firm's proposed corrective action or actions to prevent similar incidents in the future[107]

---

101. FDA, REGULATORY PROCEDURES MANUAL § 7.5.1 (2017).
102. *Id.* attachment B.
103. *Id.* attachment B, § 7.
104. *Id.* attachment B, § 1(a).
105. *Id.* attachment B, § 6.
106. *Id.* attachment B.
107. *Id.* attachment B, § 4.

- Apprise management that the district office should be consulted prior to the reconditioning or destruction of any returned product and that FDA/USDA must witness or verify product disposition[108]

In instances in which a firm has recalled a product used in the production of another product, regulatory investigators will also need the following additional information to determine whether the recall should extend to those affected products:[109]

- The firm's incoming ingredient quality control procedures[110]
- Quality control over products at the relevant time and the products in which the ingredients were used[111]
- A detailed description of methods used in preparing and packaging the product[112]
- How the finished product is stored or shipped[113]
- Labeling and cooking instructions, if any[114]
- Finished product testing information[115]
- If product is in a USDA-regulated facility, whether USDA was apprised of incoming suspect ingredients[116]

FDA's district office will also determine whether an official sample, either physical or documentary, is required.[117] Such a sample will demonstrate the product's defect and potential hazard.[118] Additionally, the samples collected should document interstate movement of the product, as well as the violation.[119] As discussed above, in both firm-initiated and agency-initiated recalls, the agency can, and often does, make extremely detailed, and often voluminous, requests for documents and information. In doing so, the agency will rely heavily on the firm's management and staff.

---

108. *Id.* attachment B, § 8.
109. OFFICE OF REGULATORY AFFAIRS, FDA, INVESTIGATIONS OPERATIONS MANUAL § 7.2.2 (2009).
110. *Id.* § 7.2.2(1).
111. *Id.* § 7.2.2(2).
112. *Id.* § 7.2.2(3).
113. *Id.* § 7.2.2(4).
114. *Id.* § 7.2.2(5).
115. *Id.* § 7.2.2(6).
116. *Id.* § 7.2.2(7).
117. FDA, *supra* note 101, § 7.5.1(3).
118. *Id.*
119. *Id.*

Admissibility of Recall in Subsequent Litigation

In large part, courts will exclude evidence of a recall if offered as proof that the recalled food was actually contaminated.[120] In fact, in some instances, evidence of a product recall may be excluded outright as a "subsequent remedial measure" under Federal Rule of Evidence 407 (or an analogous state evidence rule), although this protection is subject to some important limitations. First, courts have traditionally held that evidence of subsequent remedial measures is inadmissible to prove either negligence or culpability. The U.S. Supreme Court recognized in 1892, following an accident, that a defendant must be free to adopt safeguards "as a measure of extreme caution" without those safeguards being used as evidence against him in a civil lawsuit.[121] To admit evidence of such measures might unjustly punish responsible behavior and discourage corrective action in future cases.[122]

This long-standing principle is currently embodied in Federal Rule of Evidence 407, which reads:

> When, after an injury or harm allegedly caused by an event, measures are taken that, if taken previously, would have made the injury or harm less likely to occur, evidence of the subsequent measures is not admissible to prove negligence, culpable conduct, a defect in a product, a defect in a product's design, or a need for a warning or instruction. This rule does not require the exclusion of evidence of subsequent measures when offered for another purpose, such as proving ownership, control, or feasibility of precautionary measures, if controverted, or impeachment.[123]

Many states' evidence codes contain a provision modeled after Rule 407.[124]

---

120. *See* Thacker *ex rel.* Thacker v. Kroger Co., 155 F. App'x 946 (8th Cir. 2005).

121. Columbia & Puget Sound R.R. Co. v. Hawthorne, 144 U.S. 202 (1892) (citing Morse v. Minneapolis & St. Louis Ry., 16 N.W. 358, 359 (1883)).

122. *Id.*

123. *See* FED. R. EVID. 407 (evidence of subsequent remedial measures cannot be admitted to show liability); Bell v. Nash-Finch Co., No. 97-2191, 1999 U.S. App. LEXIS 6021 (4th Cir. Apr. 2, 1999) (under Federal Rule of Evidence 407, subsequent remedial measures are inadmissible if offered "to prove negligence, culpable conduct, a defect in a product, a defect in a product's design, or a need for a warning or instruction; thus, subsequent repairs are inadmissible to prove negligence").

124. *See, e.g.*, Alabama: ALA. R. EVID. 407 (tracking language of Federal Rule of Evidence 407); New Hampshire: N.H. R. EVID. 407 (same); Pennsylvania: PA. R. EVID. 407 (same); South Carolina: S.C. R. EVID. 407 (same).

In accordance with the rule's current requirements, courts generally consider product recalls "subsequent remedial measures," and therefore hold them to be inadmissible under Rule 407.[125] Courts analyzing recalls under Rule 407 have noted that a recall meets the rule's literal requirements because it is a (1) measure, (2) taken after an injury or harm allegedly caused by an event, (3) that, if taken previously, (4) would have made the injury or harm less likely to occur.[126] However, even before amendment made it clear that Rule 407 applied to strict product liability claims, courts justified excluding recalled evidence because the policy justifications underpinning the exclusion of post-remedial measures were strongly implicated in recall scenarios.[127] Those cases uniformly hold that recalls

---

125. *See* Chase v. Gen. Motors Corp., 856 F.2d 17 (4th Cir. 1988) (noting that recall evidence is not admissible); Bradley v. Cooper Tire & Rubber Co., No. 4:03cv00094-DPJ-JCS, 2007 U.S. Dist. LEXIS 95967 (S.D. Miss. Aug. 3, 2007) (granting motion to exclude evidence of a tire recall by Firestone); *In re* Propulsid Prods. Liab. Litig., No. 1355, 2003 U.S. Dist. LEXIS 3824 (E.D. La. Mar. 10, 2003) ("[A] recall campaign is a measure 'taken which if taken previously, would have made the event less likely to occur.' Consequently, product recalls are subsequent remedial measures for purposes of Rule 407"); Giglio v. Saab-Scania of Am., Inc., No. 90-2465, 1992 U.S. Dist. LEXIS 17026 (E.D. La. Oct. 30, 1992) (same); Bryan v. Emerson Elec. Co., 856 F.2d 192 (6th Cir. 1988) (applying Rule 407 subsequent remedial measure analysis to product recall); Thacker, 155 Fed. Appx. 946 (affirming summary judgment in favor of grocery store and ConAgra where the plaintiffs' evidence that meat was defective primarily consisted of evidence that the meat had been recalled; plaintiff failed to exclude multiple other causes of illness and to provide facts to establish proximate cause to reasonable and probable certainty); Landry v. Adam, 282 So. 2d 590 (La. Ct. App. 1973) (noting that recall evidence is not admissible); *see also* Vockie v. Gen. Motors Corp., 66 F.R.D. 57 (E.D. Pa. 1975), *aff'd*, 523 F.2d 1052 (3d Cir. 1975) (evidence of recall campaign not admissible; public policy strongly supports notices and recalls, and "[i]f such statements are admissible on a wholesale basis, manufacturers will be reluctant to come forth and make a full unqualified disclosure of any potential safety hazards which they discover. Manufacturers should not be inhibited in, or prejudiced by, a good faith effort to protect the public safety"); Buckman v. Bombadier Corp., 893 F. Supp. 547 (E.D.N.C. 1995); Hammes v. Yamaha Motor Corp. U.S.A., Inc., No. 03-6456 (MJD/JSM), 2006 U.S. Dist. LEXIS 26526 (D. Minn. May 4, 2006).

126. *Propulsid Prods. Liab. Litig.*, 2003 U.S. Dist. LEXIS 3824 ("[A] recall campaign is a measure 'taken which if taken previously, would have made the event less likely to occur.' Consequently, product recalls are subsequent remedial measures for purposes of Rule 407"); *Giglio*, 1992 U.S. Dist. LEXIS 17026 (same).

127. *Vockie*, 66 F.R.D. 57 (evidence of recall campaign not admissible; public policy strongly supports notices and recalls, and "[i]f such statements are admissible on a wholesale basis, manufacturers will be reluctant to come forth and make a full unqualified disclosure of any potential safety hazards which they discover. Manufacturers should not be inhibited in, or prejudiced by, a good faith effort to protect the

much larger than the actual problem are desirable and that companies should not be punished for taking such action as to protect the public.

Even given the protections provided by Rule 407, it is important to remember that post-remedial measures, including product recalls, are admissible for "other purposes," including "proving ownership, control, or feasibility of precautionary measures, if controverted, or impeachment."

Effectiveness Checks

Consistent with a firm's established recall strategy, the manufacturer is also responsible for creating a program for effectiveness checks to evaluate whether the consignees at the recall depth specified in the particular plan have received notification of the recall and have then taken the prescribed action.[128] These effectiveness checks can be performed through "personal visits, telephone calls, letters, or a combination thereof."[129] Although the recalling company is largely responsible for conducting these effectiveness checks, FDA will assist when necessary and appropriate.[130] The company's recall strategy will specify the method to be used and the level of effectiveness for checks that will be conducted.[131] The appropriate "effectiveness level" depends on both the scope and severity of the particular recall. "Effectiveness levels" are broken into alphabetic levels, Level A through Level E, running from the most comprehensive effectiveness check (Level A) to the least comprehensive (Level E), each correlated to a specific percentage of consignees that must be contacted.[132] The potential effectiveness levels are as follows: Level A (100 percent of the total number of consignees to be contacted);[133] Level B (some percentage of the total number of consignees to be contacted, which percentage is to be determined on a case-by-case basis, but is greater than 10 percent and less than 100 percent of the total number of consignees);[134] Level C (10 percent of the total number of consignees to be contacted);[135]

---

public safety"); *Landry*, 282 So. 2d at 597 (for reasons of public policy, evidence of or reference to recall program is improper).
  128. 21 C.F.R. § 7.42(b)(3) (2021).
  129. *Id.* (A guide entitled "Methods for Conducting Recall Effectiveness Checks" that describes the use of these different methods is available upon request from the Division of Dockets Management (HFA-305), Food and Drug Administration, 5630 Fishers Lane, rim 1061, Rockville, MD 20852.)
  130. *Id.*
  131. *Id.*
  132. *Id.*
  133. *Id.* § 7.42(b)(3)(i).
  134. *Id.* § 7.42(b)(3)(ii).
  135. *Id.* § 7.42(b)(3)(iii).

Level D (2 percent of the total number of consignees to be contacted);[136] and Level E (no effectiveness checks).[137]

Termination

Ultimately, recalls do end. Specifically:

> [A] recall will be terminated when FDA determines that all reasonable efforts have been made to remove or correct the product in accordance with the recall strategy, and when it is reasonable to assume that the product subject to the recall has been removed and proper disposition or correction has been made commensurate with the degree of hazard of the recalled product.[138]

Once a specific recall has been terminated, FDA will provide written notice of the recall termination to the recalling company.[139]

A company can request termination of its recall by submitting a written request to the appropriate FDA district office, stating that the recall is effective in accordance with the criteria set out by regulation.[140] The company's request to terminate must be accompanied by the most recent recall status report and a description of the disposition of any recalled product.[141]

---

136. *Id.* § 7.42(b)(3)(iv).
137. *Id.* § 7.42(b)(3)(v).
138. *Id.* § 7.55(a).
139. *Id.*
140. *Id.* § 7.55(b).
141. *Id.*

Chapter 6

Federal Nutrition Programs

Roger R. Szemraj and Stewart Fried

Overview

The U.S. Department of Agriculture (USDA) operates 15 nutrition assistance programs[1] administered by the Food and Nutrition Service (FNS). The largest of these programs—the Supplemental Nutrition Assistance Program (SNAP), the National School Lunch Program (NSLP), and the School Breakfast Program (SBP)—are programs with mandatory funding wherein all individuals eligible for assistance are provided that assistance. The remaining programs are discretionary funding programs wherein the number of people served and the level of assistance provided is restricted by the annual appropriations provided for these programs.[2]

As a result of the COVID-19 pandemic, the Congress has approved additional programs, discussed elsewhere in this chapter, to respond to school closures and increased unemployment.

SNAP

SNAP is the nation's largest nutrition program. Formerly known as the Food Stamp Program, SNAP provides supplemental nutrition benefits to nearly 50 million low-income Americans. SNAP's fundamental purpose

---

1. USDA Food and Nutrition Service, *Home Page*, https://www.fns.usda.gov/ (last visited May 4, 2021).

2. Congressional Budget Office, *What Is the Difference between Mandatory and Discretionary Spending?*, https://www.cbo.gov/content/what-difference-between-mandatory-and-discretionary-spending (last visited May 4, 2021).

is to provide eligible low-income households with increased food purchasing power.[3] Funding for SNAP administration and benefits is currently provided by the federal government pursuant to the 2018 Farm Bill. During fiscal year (FY) 2020, SNAP received more than $75 billion in federal appropriations, including more than $15 billion in funding related to coronavirus relief.

While SNAP is broadly administered by USDA's FNS, the enrollment of beneficiaries is conducted by state and local agencies across the country. State and local agencies generally process applications submitted by SNAP beneficiaries within 30 days, following verification of eligibility requirements.[4]

All state SNAP agencies provide benefits to eligible households via an Electronic Benefit Transfer (EBT) card. This debit-type card, which replaced the paper voucher system used for decades, permits SNAP beneficiaries to purchase eligible food items at more than 240,000 retailers authorized by FNS to redeem SNAP benefits. SNAP benefits are generally replenished monthly and, during 2020, averaged $246 in monthly household benefits.

The Food and Nutrition Act of 2008 and FNS's SNAP regulations provided few limitations on what constitutes an eligible food item. Simply put, with the exception of alcohol, tobacco, and hot foods, most food items are SNAP-eligible. Many items whose nutritional value is subject to debate are SNAP-eligible, including soda, water, energy drinks, cookies, cakes, candy, and snack foods; SNAP beneficiaries are not required to purchase staple foods with their benefits. SNAP beneficiaries are prohibited from exchanging SNAP benefits for cash or other consideration. They also may not sell food purchased with SNAP benefits or their EBT card. On the other hand, there is no prohibition on the donation or sharing of food purchased with SNAP benefits.

SNAP retailers range from small corner stores and bodegas to supermarkets and superstores. The requirements for SNAP retailer eligibility are not burdensome; retailers must carry only three varieties of foods in each of the four staple food groups: (1) meat, poultry, fish, and seafood; (2) fruits and vegetables; (3) breads, cereals, and grains; and (4) milk, cheese, and dairy products. SNAP retailers must carry only three units of each variety in each of the four staple food groups, of which two varieties must be perishable (fresh or frozen).

---

3. *See* 7 U.S.C. § 2011.
4. USDA FNS, *SNAP Eligibility*, https://www.fns.usda.gov/snap/recipient/eligibility (last updated Mar. 15, 2021).

More than 70 percent of the retail food stores that FNS authorizes to redeem SNAP benefits are convenience stores and small groceries. However, most SNAP benefits are redeemed at superstores and supermarkets. More recently, FNS has expanded the number of SNAP retailers authorized to accept SNAP benefits online, which has become especially relevant during the pandemic.[5]

FNS also authorizes a small number of entities other than retail food stores, including group homes, drug and alcohol treatment facilities, homeless and battered women's shelters, and apartment buildings primarily occupied by the elderly and disabled. In a relatively small number of authorized retailers in portions of five states, elderly, disabled, and homeless SNAP beneficiaries may redeem their benefits to purchase hot foods at otherwise ineligible businesses pursuant to the Restaurant Meals Program. Hot foods may also be SNAP-eligible in the event FNS issues a waiver, typically following weather-related disasters. SNAP retailers also avoid the prohibition on the sale of hot foods by selling foods cold to SNAP beneficiaries and cooking or heating them post-sale; such activity is permissible as long as SNAP benefits are not used to pay for any cooking or heating charges and if less than 50 percent of the retailer's gross receipts consist of foods that are cooked, heated, or prepared on-site, presale or post-sale.[6]

FNS aggressively combats SNAP fraud through a variety of tools and programs. The agency frequently sends undercover investigators to smaller SNAP retailers to verify compliance with SNAP regulations. These covert efforts often involve the purchase of common nonfood items with SNAP benefits, including paper towels, toilet paper, and cleaning supplies. The penalty for the sale of common nonfoods in exchange for SNAP benefits is a six-month term disqualification for a first violation and a year for a second violation.[7] The penalty for the sale of alcohol or tobacco products is five years if the store had been previously warned that violations may be occurring.[8]

The most serious penalty that a SNAP retailer may face is permanent disqualification for trafficking in SNAP benefits.[9] The most frequent form of "trafficking" is the exchange of cash or other consideration for

---

5. USDA FNS, *FNS Launches the Online Purchasing Pilot*, https://www.fns.usda.gov/snap/online-purchasing-pilot (last updated Apr. 27, 2021).
6. 7 C.F.R. § 278.1(b)(iv).
7. *Id.* § 278.6(e)(5) and (6).
8. *Id.* § 278.6(e)(2).
9. *Id.* § 278.6(e)(1).

SNAP benefits.[10] FNS charges SNAP retailers with trafficking in two primary ways: (1) pursuant to an undercover investigation in which a clerk or owner gives the undercover investigator cash from the register in exchange for the redemption of a higher amount of SNAP benefits; and (2) based on patterns of SNAP transactions that the agency deems suspicious, unusual, or inexplicable. While combatting SNAP fraud is an appropriate goal, FNS's use of its Anti-Fraud Locator Using the Electronic Benefits Transfer Retailer Transactions (ALERT) system is troubling for numerous reasons, including the lack of any regulations or guidance related thereto. Unfortunately, it is altogether unclear what, if anything, FNS has done to establish that the ALERT system patterns are reliable or if it has conducted any audits or taken any other action to ensure that SNAP retailers are not improperly disqualified. Given that revenues from SNAP beneficiaries' purchases often make up a significant portion of corner stores' or small grocers' gross receipts, especially those in low-income neighborhoods, getting disqualified often has material effects on these businesses' operations.

FNS's administrative review process is also a serious concern for small retailers. Those proceedings are not contested cases before an impartial administrative law judge; rather, SNAP administrative review proceedings are decided by administrative review officers (AROs), who are not lawyers. No opportunity for a hearing is available and SNAP retailers do not have the ability to conduct discovery. Prior to October 2020, SNAP retailers had the ability to request records pursuant to a Freedom of Information Act (FOIA) request; and while the agency often utilized FOIA exemptions to prevent the release of many records used by FNS officials in deciding whether to impose a disqualification or not, retailers did receive some records other than the letter charging them with trafficking and other violations.

Effective October 2020, FNS instituted a new regulation that effectively precludes SNAP retailers from using FOIA to attempt to discover records in the agency's possession relating to their SNAP authorization and FNS's charges; this new regulation prohibits the agency from holding administrative proceedings in abeyance during the pendency of SNAP administrative proceedings.[11] Ostensibly, the agency's rationale was to prevent SNAP retailers from delaying administrative proceedings during

---

10. *Id.* § 271.2.
11. *See* Taking Administrative Actions Pending Freedom of Information Act (FOIA) Processing, 85 Fed. Reg. 52,471 (Aug. 26, 2020), *available at* https://www.govinfo.gov/content/pkg/FR-2020-08-26/pdf/2020-18701.pdf.

the pendency of the FOIA process. The effect of FNS's regulatory change is that SNAP retailers are not afforded access to any of the key records that AROs and other FNS officials will review in deciding whether to permanently disqualify the retailer. This administrative construct is unique among USDA agencies and has been described as Kafkaesque and pejoratively compared with Star Chamber proceedings. Not surprisingly, FNS permanently disqualifies nearly all SNAP retailers it charges with trafficking and other program violations. SNAP retailers fare little better on administrative review; FNS's Administrative Review Branch reverses less than 1 percent of SNAP administrative appeals. The playing field changes dramatically if disqualified retailers seek judicial review; there, aggrieved retailers are provided access to the entire administrative record, typically pursuant to a protective order. Judicial review proceedings are reviewed by judges under a de novo standard and discovery is generally permitted.[12]

In sum, SNAP offers tremendous benefits to beneficiaries and retailers. While the majority of SNAP dollars are spent at large supermarkets and superstores, small retailers and corner stores constitute the majority of individually authorized retail food stores that participate in SNAP. In pro bono practice, lawyers have an opportunity to assist individual SNAP beneficiaries via pro bono administrative appeals and public benefits outreach. Lawyers also have an opportunity to assist retailers of all sizes participating in SNAP to assess compliance risks, address issues identified by FNS, and handle administrative proceedings before the agency and on judicial review.

Child Nutrition Programs

The primary child nutrition programs are the NSLP, authorized by the Richard B. Russell National School Lunch Act (the Russell Act),[13] and the SBP, authorized by section 4 of the Child Nutrition Act (CNA).[14] These programs were reauthorized by the Healthy, Hunger-Free Kids Act of 2010 (HHFKA),[15] and have continued under annual appropriation bills.

Since the adoption of the Russell Act in 1946, the stated policy has been

---

12. 7 U.S.C. § 2023(a)(15).
13. 42 U.S.C. § 1751 *et seq.*
14. *Id.* § 1773.
15. Pub. L. No. 111-296, 124 Stat. 3183.

as a measure of national security, to safeguard the health and well-being of the Nation's children and to encourage the domestic consumption of nutritious agricultural commodities and other food, by assisting the States, through grants-in-aid and other means, in providing an adequate supply of foods and other facilities for the establishment, maintenance, operation, and expansion of nonprofit school lunch programs.

Food vendors, service contractors, and food service management companies (collectively referred to here as "contractors") make their contracts with state agencies (SAs), local education agencies (LEAs), or school food authorities (SFAs), as appropriate. The procurement terms are set by those agencies and authorities, and are beyond the scope of this discussion. However, since these entities and their operations need to be in compliance with federal requirements as specified by regulation and statutes, contractors routinely find that their own compliance with the federal requirements is a condition of the procurement and ultimate award.

The NSLP and SBP are offered in public and nonprofit elementary and secondary schools, including charter schools. Schools are not required to participate in the programs. The programs are administered by SAs that have agreements with participating schools. Those schools are provided support with cash subsidies for each meal served. Meal rates are provided for contiguous states, and modified for Alaska, Guam, Hawaii, Puerto Rico, and the Virgin Islands. Rates are set for those schools serving less than 60 percent free and reduced-price meals, and these rates are increased for those serving 60 percent or more. SFAs that have been certified by their SA to meet the meal pattern and nutrition standards set forth in 7 C.F.R. parts 210 and 220 will receive an additional seven cents.[16] For school year (SY) 2020–2021, the primary lunch rate including the additional seven cents is $3.58 per free lunch, and $3.18 per reduced-price lunch, while the free breakfast rate is $1.89 and the reduced breakfast rate is $1.59.[17] Those participating in the NSLP also receive a commodity entitlement for the period July 1, 2020, through

---

16. Memorandum SP 26-2014 from USDA FNS, to Regional Directors, Special Nutrition Programs, and State Directors, Child Nutrition Programs, Re: New Questions and Answers Related to the Certification of Compliance with Meal Requirements for the National School Lunch Program (Mar. 11, 2014).

17. National School Lunch, Special Milk, and School Breakfast Programs, National Average Payments/Maximum Reimbursement Rates, 85 Fed. Reg. 44,270 (July 22, 2020).

June 30, 2021, of 24.50 cents per meal, which can be used to purchase USDA Foods.[18] Commodity entitlements are determined based on the estimated number of lunches served between July 1 and June 30 of the preceding SY in schools participating in the NSLP.[19] USDA Foods are discussed in detail below.

According to FNS, in FY 2019 more than 4.8 billion lunches were served to an average of nearly 30 million children each day.[20] More than 74 percent of these meals were served to children participating in the free or reduced-price programs. During the same period, nearly 2.5 billion breakfasts were served to an average of nearly 14.8 million children each day, with more than 85 percent of breakfasts served to children participating in the free or reduced-price programs.[21]

How Children Qualify for Free and Reduced-Price Meals

Families are required to submit applications to their SFAs for their children to participate in the free or reduced-price NSLP or SBP. Children from families with incomes at or below 130 percent of the federal poverty level are eligible for free meals. Those with incomes between 130 and 185 percent of the federal poverty level are eligible for reduced-price meals. Eligible family income levels are based on household size. For SY 2020–2021,[22] children from a family of four in the 48 contiguous states, the District of Columbia, Guam, and the territories with an annual income of $34,060 or less are eligible for free meals. Those with incomes between $34,060 and $48,470 are eligible for reduced-price meals. Eligible income levels for those living in Alaska or Hawaii are higher. Schools can charge a maximum of 40 cents for a reduced-price lunch.

Children are also "categorically eligible" and are directly certified for free meals if they participate in other federal assistance programs including SNAP, the Food Distribution Program on Indian Reservations

---

18. Food Distribution Program: Value of Donated Foods from July 1, 2020 through June 30, 2021, 85 Fed. Reg. 44,273 (July 22, 2020).

19. 42 U.S.C. § 1755(c).

20. FNS, USDA, NATIONAL SCHOOL LUNCH PROGRAM: PARTICIPATION AND LUNCHES SERVED (2021), https://fns-prod.azureedge.net/sites/default/files/resource-files/slsummar-4.pdf.

21. FNS, USDA, SCHOOL BREAKFAST PROGRAM PARTICIPATION AND MEALS SERVED (2021), https://fns-prod.azureedge.net/sites/default/files/resource-files/sbsummmar-4.pdf.

22. Child Nutrition Programs: Income Eligibility Guidelines, 55 Fed. Reg. 16,050 (Mar. 20, 2020).

(FDPIR), or Temporary Assistance for Needy Families (TANF).[23] They may also be eligible based on their status as a homeless, migrant, runaway, or foster child.[24]

Some SFAs and LEAs utilize the Community Eligibility Provision (CEP) to avoid the application process. CEP is available if at least 40 percent of students in the prior SY were directly certified through SNAP, FDPIR, or TANF, or were certified for free school meals without an application because they are homeless, migrant, runaway, enrolled in Head Start, or in foster care.[25] A formula is used to determine how much federal reimbursement will be provided for the meals served in CEP schools, and the local authority must provide nonfederal funds to cover costs above USDA reimbursement.[26]

There are also special alternatives for annual determinations of eligibility for free and reduced-price school meals and daily meal counts by type (free, reduced-price, and paid meals) at the point of service.[27] According to FNS,[28] the alternatives are:

- **Provision 1**. This allows free eligibility to be certified for a two-year period. In schools where at least 80 percent of the children enrolled are eligible for free or reduced-price meals, annual notification of program availability and certification of children eligible for free meals may be reduced to once every two consecutive SYs.
- **Provision 2**. This allows schools to establish claiming percentages and to serve all meals at no charge for a four-year period. Schools must offer meals to all participating children at no charge for a period of four years.
- **Provision 3**. This allows schools to simply receive the same level of federal cash and commodity assistance each year, with some adjustments, for a four-year period. Schools must serve meals to all participating children at no charge for a period of four years.

---

23. 7 C.F.R. § 245.6(b)(7).
24. *Id.* § 245.6(b)(8).
25. *Id.* § 245.9(f).
26. USDA FNS, *Community Eligibility Provision Resource Center*, https://www.fns.usda.gov/nslp/community-eligibility-provision-resource-center (last updated Aug. 8, 2020).
27. 42 U.S.C. § 1759a(a)(1).
28. USDA FNS, *Provisions 1, 2, and 3*, https://www.fns.usda.gov/school-meals/provisions-1-2-and-3 (last updated May 6, 2014).

What Meals May Be Served

Meals are designed by SFAs. But a school lunch or breakfast is eligible for federal reimbursement only if it meets the nutrition standards specified by regulation.[29] The NSLP requires five food components, each with daily and weekly minimums, including fruits, vegetables, grains, meats/meat alternates, and fluid milk. The SBP requires three food components, each with daily and weekly minimums, including fruits (with some allowance for vegetable substitution), grains (with some allowance for meat/meat alternate substitution), and fluid milk.[30]

Following the enactment of HHFKA in 2010, significant revisions to school meal nutrition standards were implemented in 2012 following a comment period in which USDA received a total of 133,268 public comments including 7,107 unique letters, 122,715 form letters from 159 mass mail campaigns, 3,353 nongermane letters, and 93 duplicates.[31] Due to complaints from SFAs and others regarding difficulty in complying with portions of these standards, the Congress included provisions in the FY 2015, FY 2016, and FY 2017 appropriations acts that eased milk, whole-grain, and/or sodium requirements from SY 2015–2016 through SY 2017–2018.[32]

HHFKA also required that the Secretary of Agriculture issue regulations to establish nutrition standards for foods sold in schools other than those foods provided under the Russell Act and CNA.[33] These "competitive foods" includes all food and beverages other than meals reimbursed under programs authorized by the Russell Act and CNA available for sale to students on the school campus during the school day.

---

29. 7 C.F.R. §§ 210.10, 220.8; USDA FNS, *Nutrition Standards for School Meals*, https://www.fns.usda.gov/school-meals/nutrition-standards-school-meals (last updated Sept. 23, 2019).

30. Memorandum SP38-2019 from USDA FNS, to Regional Directors, Special Nutrition Programs, and State Directors, Child Nutrition Programs, Re: Meal Requirements under the National School Lunch Program and School Breakfast Program: Questions and Answers for Program Operators (Sept. 23, 2019).

31. Nutrition Standards in the National School Lunch and School Breakfast Programs, 77 Fed. Reg. 4,088 (Jan. 26, 2012).

32. Pub. L. No. 113-235, §§ 751 and 752, 128 Stat. 2130 (2014); Pub. L. No. 114-113, § 733, 129 Stat. 2242 (2015); Pub. L. No. 115-31, § 747, 131 Stat. 568 (2017).

33. National School Lunch Program and School Breakfast Program: Nutrition Standards for All Foods Sold in School as Required by the Healthy, Hunger-Free Kids Act of 2010—Interim Final Rule, 78 Fed. Reg. 39,068 (June 28, 2013); Final Rule, 81 Fed. Reg. 50,131 (July 29, 2016).

Buy American Requirement

The Russell Act provides that "the Secretary shall require that a school food authority purchase, to the maximum extent practicable, domestic commodities or products."[34] The Russell Act defines "domestic commodity or product" as "(A) an agricultural commodity that is produced in the United States; and (B) a food product that is processed in the United States substantially using agricultural commodities that are produced in the United States." SFAs are expected to comply with the Buy American requirement when providing a meal eligible for reimbursement, and to be prepared to document any exceptions to this requirement.

FNS has issued policy memoranda regarding the Buy American provisions that are worthy of note.[35] These memoranda emphasize that in order to be compliant with the Buy American requirements, agricultural commodities and the products of agricultural commodities shall be processed in the United States and substantially contain meats, vegetables, fruits, fish, and other agricultural commodities produced in the United States or any territory or possession of the United States.

State Administrative Expenses

The base amount of state administrative expenses available for allocation to states is equal to at least 1.5 percent of federal cash program payments for the NSLP (excluding snacks), SBP, Child and Adult Care Food Program (CACFP) (including snacks), and Special Milk Program in the second previous FY (i.e., FY 2018 for FY 2020). State administrative expense funds are used to support state employee salaries, benefits, and other allowable administrative expenses necessary to implement and oversee program operations.

COVID-19 Activities

The Congress has provided a variety of waivers for child nutrition programs as part of COVID-19 relief legislation. Those waivers have enabled children to continue to receive food assistance either with direct

---

34. 42 U.S.C. § 1760(n); 7 C.F.R. § 210.21(d).
35. Memorandum SP 38-2017 from USDA FNS, to Regional Directors, Special Nutrition Programs, and State Directors, Child Nutrition Programs, Re: Compliance with and Enforcement of the Buy American Provision in the National School Lunch Program (June 30, 2017); Memorandum SP 32-2019 from USDA FNS, to Regional Directors, Special Nutrition Programs, and State Directors, Child Nutrition Programs, Re: Buy American and the Agriculture Improvement Act of 2018 (Aug. 15, 2019).

distribution or through a Pandemic EBT program with SNAP benefits.[36] The continuation of these waivers and Pandemic EBT in SY 2020–2021 will be important and will likely have implications for SY 2021–2022. SFAs and advocacy groups continue to suggest that some of these waivers may be needed in the next SY and perhaps beyond as school meal participation builds back up to prior levels.

Summer Food Service Program

The Summer Food Service Program provides reimbursement to those who serve free, healthy meals to children and teens who are 18 years old or younger in low-income areas during the summer months when school is not in session. The program is administered by SAs that enter into agreements with providers. The providers are usually schools, local governments, camps, and faith-based and other nonprofit community organizations who offer meals at central locations.[37]

CACFP

CACFP[38] provides reimbursements for meals and snacks (shown as "supplements" in the table below) to eligible children and adults who are enrolled for care at participating child care centers or day care homes; children and youth participating in afterschool care programs; children residing in emergency shelters; adults at adult day care centers; and adults over the age of 60 or living with a disability and enrolled in day care facilities. The program is administered by SAs.

Eligibility

Those receiving program benefits must meet the income guidelines issued annually.[39] In centers, participants from households with incomes at or below 130 percent of the federal poverty level are eligible for free meals while those with household incomes between 130 percent and 185 percent of the federal poverty level are eligible for meals at a reduced price. The program provider is responsible for determining income eligibility.

---

36. United States Department of Agriculture, Food and Nutrition Service. "Pandemic EBT (P-EBT) Questions and Answers" (April 26, 2021).
37. 7 C.F.R. pt. 225.
38. *Id.* pt. 226.
39. Child Nutrition Programs: Income Eligibility Guidelines, 55 Fed. Reg. 16,050 (Mar. 20, 2020).

Householding organizations of day care homes must determine which day care homes are eligible for tier I rates and, if requested, which children are eligible to receive meals reimbursed at tier I rates in tier II day care homes.

Automatic eligibility is provided for children participating in Head Start or Even Start, for foster children who are the responsibility of the state or placed by the court, for homeless children, and for adults who receive SNAP, FDPIR, Social Security Income, or Medicaid benefits.

Reimbursement Rates

Reimbursement rates for meals are adjusted annually and are shown in the following table.[40] USDA supplements these rates with donated agricultural foods or cash-in-lieu of donated foods available to institutions participating in CACFP.

CACFP
(Per meal rates in whole or fractions of U.S. dollars, effective from July 1, 2020, to June 30, 2021)

Centers	Breakfast	Lunch and Supper*	Supplement
Contiguous States:			
Paid	0.32	0.33	0.08
Reduced-Price	1.59	3.11	0.48
Free	1.89	3.51	0.96
Alaska:			
Paid	0.49	0.54	0.14
Reduced-Price	2.73	5.30	0.78
Free	3.03	5.70	1.56
Hawaii:			
Paid	0.37	0.39	0.10
Reduced-Price	1.91	3.71	0.56
Free	2.21	4.11	1.13

---

40. Child and Adult Care Food Program: National Average Payment Rates, Day Care Home Food Service Payment Rates, and Administrative Reimbursement Rates for Sponsoring Organizations of Day Care Homes for the Period July 1, 2020 through June 30, 2021, 85 Fed. Reg. 44,268 (July 22, 2020).

Day Care Homes	Breakfast		Lunch and Supper		Supplement	
	Tier I	Tier II	Tier I	Tier II	Tier I	Tier II
Contiguous States	1.39	0.50	2.61	1.58	0.78	0.21
Alaska	2.22	0.78	4.24	2.55	1.26	0.35
Hawaii	1.62	0.58	3.06	1.84	0.91	0.25
Administrative reimbursement rates for sponsoring organizations of day care homes *(per home/per month rates in U.S. dollars)*			Initial 50	Next 150	Next 800	Each Additional
Contiguous States			120	91	71	63
Alaska			194	148	116	102
Hawaii			140	107	84	73

* These rates do not include the value of USDA Foods or cash-in-lieu of USDA Foods that institutions receive as additional assistance for each CACFP lunch or supper served to participants. A notice announcing the value of USDA Foods and cash-in-lieu of USDA Foods is published separately in the *Federal Register*.

Special Supplemental Nutrition Program for Women, Infants, and Children

The Special Supplemental Nutrition Program for Women, Infants, and Children (WIC), established in 1974, provides grants to states for supplemental foods, health care referrals, and nutrition education for low-income pregnant, breastfeeding, and non-breastfeeding postpartum women, and to infants and children up to age five who are found to be at nutritional risk. It is the largest discretionary spending food assistance program offered by FNS. But, while it is discretionary, it has been consistently funded at a level sufficient to serve applicants. WIC is authorized by section 17 of the CNA.[41]

According to FNS, WIC is administered by 90 SAs, through approximately 47,000 authorized retailers. WIC operates through 1,900 local agencies in 10,000 clinic sites, in 50 state health departments, 34 Indian tribal organizations, the District of Columbia, and five territories (Northern Marianas, American Samoa, Guam, Puerto Rico, and the Virgin Islands). To be eligible for the program, the applicant's gross income must be at or below 185 percent of the U.S. Poverty Income Guidelines,

---

41. 42 U.S.C. § 1786; 7 C.F.R. pt. 246.

currently $48,470 for a family of four. The applicant must also be individually determined to be at "nutritional risk" by a health professional or a trained health official.

The Food Package
The monthly WIC food package is defined based upon the age and status of the individual participant. It will include a combination of infant cereal, iron-fortified adult cereal, vitamin C-rich fruit or vegetable juice, eggs, milk, cheese, peanut butter, dried and canned beans/peas, canned fish, soy-based beverages, tofu, fruits and vegetables, baby foods, whole wheat bread, and other whole-grain options.[42]

While breastfeeding is strongly encouraged in WIC, infant formula is provided by SAs for mothers who choose to use formula. These SAs are required by law to have competitively bid infant formula rebate contracts with infant formula manufacturers.[43] The SA agrees to provide only one brand of infant formula, and the brand will vary from each state depending upon which manufacturer wins the contract. The manufacturer gives the SA a rebate for each can of infant formula purchased by WIC participants.

Where WIC Benefits Can Be Redeemed
WIC participants normally redeem their benefits at authorized grocery stores. There are limited instances (i.e. remote locations) in which the food package may be delivered to the participant. SAs have been ordered to provide statewide EBT by October 1, 2020.

In order to redeem WIC benefits, the retailer must be authorized by the responsible SA.[44]

The Emergency Food Assistance Program

The Emergency Food Assistance Program (TEFAP) provides USDA Foods (discussed below) for distribution to eligible individuals through state distributing agencies (SDAs).[45] SDAs approve local organizations such as food banks who distribute the food with monthly distributions to eligible individuals.

---

42. 7 C.F.R. § 246.10(e).
43. 42 U.S.C. § 1786(h)(8); 7 C.F.R. § 246.16a.
44. WIC SAs may be found at USDA FNS, *Contacts*, https://www.fns.usda.gov/contacts?f%5B0%5D=program%3A32 (last visited May 4, 2021).
45. 7 U.S.C. § 7501 note; 7 C.F.R. pt. 251.

What Foods Are Available

USDA currently makes available more than 120 products, including canned, frozen, dried, and fresh fruits and vegetables; eggs; meat; poultry; fish; nuts; milk and cheese; and whole-grain and enriched grain products including rice, cereal, and pasta. The individual products available in each state will vary based upon local preferences and the response to product solicitations issued by the Agricultural Marketing Service (AMS) as directed by FNS.[46] Foods are purchased each year using funds designated for TEFAP commodities as part of the mandatory SNAP appropriation. These foods are often supplemented with the purchase of other commodities acquired as "bonus items" under USDA's market support authorities.

What Operating Assistance Is Available

USDA also provides TEFAP administrative funding to states to support the storage and distribution of USDA Foods and foods from other sources, including private donations. This funding is provided through annual discretionary appropriations. States are allowed to convert up to 15 percent of TEFAP food funds to administrative funds. States may also convert any amount of their administrative funds to food funds to purchase additional USDA Foods, and to support gleaning and other food recovery efforts.

Who Is an Eligible Recipient?

SAs set uniform criteria for eligibility to receive commodities. The criteria must (1) enable the SA to ensure that only households that are in need of food assistance because of inadequate household income receive TEFAP commodities; (2) include income-based standards and the methods by which households may demonstrate eligibility under such standards; and (3) include a requirement that the household reside in the geographic location served by the SA at the time of applying for assistance, but length of residency shall not be used as an eligibility criterion.[47]

COVID-19 Response

Additional appropriations have been provided to supplement both the ability to purchase TEFAP foods, and TEFAP operating expenses. The Farmers to Families Food Box Program has also provided an additional

---

46. USDA FNS, *USDA Foods Available List for TEFAP*, https://www.fns.usda.gov/tefap/usda-foods-available-list-tefap (last updated Mar. 15, 2021).

47. 7 C.F.R. § 251.5(b).

way to provide certain meats, fruits and vegetables, and certain dairy products to those in need.[48]

Commodity Supplemental Food Program

The Commodity Supplemental Food Program (CSFP)[49] is a discretionary program that provides monthly USDA Foods[50] to persons at least 60 years of age with incomes at or below 130 percent of the federal poverty level.[51] CSFP is available in all 50 states, the District of Columbia, Puerto Rico, and through the following Indian tribal organizations: Oglala Sioux (South Dakota), Red Lake (Minnesota), Seminole Nation (Oklahoma), Shingle Springs Band of Miwok Indians (California), and Spirit Lake Sioux Tribe (North Dakota). Yet, for the most part, it operates in only portions of each state as determined by state plans filed with FNS. The local service agency is often a food bank. State caseloads are awarded each year based upon available funds, and the total caseload was 736,110 for the 2020 caseload cycle.[52] For FY 2020, an administrative grant of $81.15 was provided. The grant is adjusted each year for inflation.

USDA Foods

The USDA Foods Program provides 100 percent American-grown and -produced foods for use in schools, TEFAP, CSFP, FDPIR, and emergency relief. The more than 200 foods available vary by program and are made available in institutional and household sizes, as may be appropriate.[53]

---

48. *See* USDA AMS, *USDA Farmers to Families Food Box*, https://www.ams.usda.gov/selling-food-to-usda/farmers-to-families-food-box (last visited May 4, 2021).

49. 7 U.S.C. § 612c note; 7 C.F.R. pts. 247 and 250.

50. CSFP foods available can be found at USDA FNS, *USDA Foods Available List for CSFP 2021*, https://www.fns.usda.gov/csfp/csfp-foods-available (last updated Sept. 2, 2019).

51. 2020 income guidelines can be found at USDA FNS, *CSFP: Income Guidelines for 2020*, https://www.fns.usda.gov/csfp/income-guidelines-2020 (last updated Feb. 11, 2020).

52. 2020 caseloads and administrative grants can be found at USDA FNS, *CSFP: Final Caseload Assignments for the 2020 Caseload Cycle and Administrative Grants*, https://www.fns.usda.gov/csfp/final-caseload-assignments-2020-caseload-cycle (last updated Jan. 9, 2020).

53. The foods available lists for each program may be found at USDA FNS, *USDA Foods Expected to Be Available*, https://www.fns.usda.gov/usda-foods/usda-foods-expected-be-available (last updated Dec. 16, 2020).

FNS provides specifications for each food item.[54] The AMS procures food items as requested by FNS through competitively bid solicitations throughout each year.[55] Only those vendors who have been approved by AMS may submit bids.[56]

AMS may include small business and service-disabled veteran-owned small business set-asides as part of these various procurements.

AMS from time to time may conduct audits and inspections of approved vendors for contract compliance at the cost of the contractor.

AMS also operates the Food Purchase and Distribution Program as part of trade mitigation efforts to assist producers impacted by trade retaliation by foreign nations.

Dietary Guidelines for Americans

The Dietary Guidelines for Americans (the Guidelines) report issued jointly by USDA and the U.S. Department of Health and Human Services (HHS) is required every five years.[57] FNS is the USDA agency responsible for leading the effort for the year ending in "0," while the *Guidelines* are to be based on the preponderance of current scientific and medical knowledge. The Guidelines are intended to provide Americans with science-based recommendations for the healthy consumption of food.

The 2020 edition of the Guidelines[58] for the first time includes specific recommendations for those from birth to 24 months and for people who are pregnant.

---

54. Food product specifications may be found at USDA AMS, *Product Specifications & Requirements*, https://www.ams.usda.gov/selling-food/product-specs (last visited May 4, 2021).

55. Master solicitations and solicitations schedules may be found at USDA AMS, *Purchase Programs: Solicitations & Awards*, https://www.ams.usda.gov/selling-food/solicitations (last visited May 4, 2021).

56. *See* Webinar: How to Become a Certified USDA Vendor (USDA AMS 2018), https://www.youtube.com/watch?v=i_e36KkRDzo; AMS, USDA, NEW VENDOR QUALIFICATION CHECKLIST (2018), https://www.ams.usda.gov/sites/default/files/media/NewVendorQualificationChecklist.pdf.

57. 7 U.S.C. § 5341 *et seq.*

58. USDA, DIETARY GUIDELINES FOR AMERICANS 2020–2025 (2020), https://www.dietaryguidelines.gov/sites/default/files/2020-12/Dietary_Guidelines_for_Americans_2020-2025.pdf.

Process

The process for developing the updated Guidelines begins with the appointment of a Dietary Guidelines Advisory Committee (the Committee) composed of nutrition and medical researchers, academics, and practitioners selected by HHS and USDA. The Committee reviews current scientific information, conducts public hearings, and solicits public comments before issuing an advisory report containing the Committee's recommendations. This advisory report is then open to public comment before final approval and issuance of the updated Guidelines by HHS and USDA.

Importance of the Guidelines to Nutrition Programs

The Guidelines are key elements for the nutrition standards in multiple nutrition programs.

The Russell Act requires that "the Secretary shall promulgate rules, based on the most recent Dietary Guidelines for Americans, that reflect specific recommendations, expressed in serving recommendations, for increased consumption of foods and food ingredients offered in school nutrition programs under this Act and the Child Nutrition Act of 1966."[59] Schools participating in the lunch and breakfast programs are required to "serve lunches and breakfasts that . . . are consistent with the goals of the most recent Dietary Guidelines for Americans."[60]

CACFP requires that "[n]ot less frequently than once every 10 years, the Secretary shall review and, as appropriate, update requirements for meals served under the program under this section to ensure that the meals . . . are consistent with the goals of the most recent Dietary Guidelines."[61]

The WIC program is also aligned with the Guidelines.[62]

Opportunities for Involvement

There are multiple opportunities for involvement by attorneys. The legislative process for the reauthorization for child nutrition programs is expected to get underway in 2021 in the Senate Agriculture Committee

---

59. 42 U.S.C. § 1758(a)(4)(B).
60. *Id.* § 1758(f)(1)(A).
61. *Id.* § 1766(g)(2)(B)(i).
62. Special Supplemental Nutrition Program for Women, Infants, and Children (WIC): Revisions in the WIC Food Packages; Interim Rule, 72 Fed. Reg. 68,966 (Dec. 6, 2007).

and the House Education and Labor Committee. Work by the House and Senate Agriculture Committees on the next Farm Bill, which includes reauthorization for SNAP, TEFAP, CSFP, and other food purchase programs, may also begin later this year.

USDA is also in the process of reviewing regulatory requirements for multiple food assistance programs, which will likely offer opportunities for public comment.

There are also ongoing contracting actions undertaken by SAs, SFAs, and USDA itself for the provision of food items and related services. Compliance with federal regulations as well as state and local requirements are key elements of a company's ability to successfully provide the solicited items.

Chapter 7

Practicing International Food Law: Considerations for Counsel

Charles F. Woodhouse

Is There Such a Thing as "International Food Law"?

Asking a practicing attorney who specializes in "food law" what is "international food law" is probably not going to lead to a useful answer. My law practice is entirely "food law" and substantially "international," yet I cannot say that there even is such a thing as "international food law" that can be clearly defined and categorized. This is of course because the international system is not one body of law, but many overlapping sources of customs, norms, policies, and national rules.

When faced with the prospect of an impossibly complex definition of "international food law," the easy way to get closer to a definition is to sketch it out, and we have done so in the following table.

Sources of International Food Law Norms	
• Compliance • Consumer disclosure • National norms • National and ethnic identity norms • Permissible marketing claims • International food safety standards • International trade standards and World Trade Organization (WTO) treaties	• Supply chain transport and logistics • Manufacturer/distributor liability and insurance • Customs compliance • National food labelling compliance including allergen disclosure • Good Manufacturing Practices (GMPs) and Good Agricultural Practices (GAPs) • Cultural factors and traditions including trademark law restrictions • National law • *Codex Alimentarius* and WTO SPS and TBT treaties

Over my years in law practice, I have seen international food law best conceptualized as encompassing all those issues and problems related to the production and distribution of human and animal foods within the United States and the movement of food final products and ingredients into and out of the United States. The international movement of food products and ingredients involves, of course, being able to deal with regulatory issues in both the country of origin and the destination jurisdiction.

Naturally, no one of us can possibly "master" each of these diverse topics—so what can we do as lawyers? The answer that can best guide us was given by Associate Justice Oliver Wendell Holmes, Jr., more than a century ago:

> When we study law we are not studying a mystery but a well known profession. We are studying what we shall want in order to appear before judges, or to advise people in such a way as to keep them out of court. The reason why it is a profession, why people will pay lawyers to argue for them or to advise them, is that in societies like ours the command of the public force is intrusted [sic] to the judges in certain cases, and the whole power of the state will be put forth, if necessary, to carry out their judgments and decrees. People want to know under what circumstances and how far they will run the risk of coming against what is so much stronger than themselves, and hence it becomes a business to find out when this danger is to be feared. The object of our study, then, is prediction, the prediction of the incidence of the public force through the instrumentality of the courts.[1]

---

1. Oliver Wendell Holmes, Jr., *The Path of the Law*, 10 HARV. L. REV. 457, 457 (1897).

The business of food production and distribution is highly regulated in every country and often by local governments within the modern nation state. Thus, the business entity engaged in this business is subject to Holmes's "public force" in every country. Thus, our task as "food and ag lawyers" is to guide clients in compliance and to "predict" for our clients the areas in which they may come up against this public force that is manifested in recalls, import refusals, labeling regulations, and a myriad of other regulatory actions of government.

Getting Started in International Food Law: Education

Before we move into a detailed discussion of the sources of law described in the above table, we will briefly comment on a topic of fundamental importance—the education and training of a person who aspires to be an "international food lawyer" or perhaps preferably a person who is a food lawyer but who needs to be able to deal with international regulatory issues. This topic will comprise a major portion of this chapter, as we discuss training and education in detail below.

Since international food law is too broad a topic for any one individual, some relevant details of the author's background are presented to acknowledge his own limitations. My background involves undergraduate pre-med at Dartmouth; an M.B.A. at Wharton (thesis on commodities futures trading); a Rutgers J.D. with the senior research topic of vicarious liability of corporate actors in criminal enforcement of the Federal Food, Drug, and Cosmetic Act; M.S. in food safety from Michigan State University (food chemistry, microbiology, toxicology); M.S. in packaging from Michigan State (food engineering and food contact materials for the preservation and protection of food); and I am currently a graduate student at the University of Florida's Institute of Food and Agricultural Sciences, where my research agenda focuses on the contamination, primarily through agricultural irrigation water, of ready-to-eat (RTE) foods with pathogens and natural and anthropogenic poisons.

My particular focus is food safety (including packaging) for RTE Foods. "RTE" of course is "ready-to-eat" and represents the highest category of danger to human health since there is no "kill step" at the end of the line when the consumer is served the food in a restaurant or brings it home from the retail grocer. RTE foods include dairy, many seafood items such as smoked fish, preserved meats, and many produce items. We will discuss in detail the specific Good Management Practice (GMP) issues surrounding RTE foods later in this chapter. After considerable delay, the U.S. Food and Drug Administration (FDA) has issued a proposed rule that lists these high-risk foods.[2] These topics are not merely of academic interest but are essential in the professional preparation of Risk-Based Preventive Control (RBPC) plans, which are mandatory documents for food producers and processors under the rules promulgated pursuant to the Food Safety Modernization Act (FSMA) and followed, in whole or in part, by most modern countries.

As a result, the focus of my work often regards ensuring that food remains safe to eat along its journey from the farm to the fork, even when that journey takes the food product across national boundaries.

So, how should we view "international food law" in order to collect all its pieces into a single collective context for discussion? We will start with a brief explanation of the issues and then let the reader's curiosity and imagination take her or him further into the complexities that will require years of study.

The Truly "International" Food Regulatory System

This is the area of certain international institutions and organizations established through international treaties that are part of the system of "public international law." Many of us may have taken the standard law school course on this subject—as opposed to "private international law" (termed "conflict of laws" at most American law schools).

As a foundation to work in international food law, it is often important for a practitioner of international food law to have a solid grounding in the work of the United Nations Food and Agriculture Organization (FAO), the World Health Organization (WHO), and the World Trade Organization (WTO).

Every modern developed nation's food safety system relies heavily upon the food standards established by the *Codex Alimentarius* Commission (FAO-WHO). In fact, the *Codex Alimentarius* standards are often the basis of domestic standards established by the U.S. Department of Agriculture (USDA), FDA, the European Commission, and Health Canada.

This is so important that major research universities, such as Michigan State, offer full-semester graduate courses on the *Codex Alimentarius* standards, which are so ubiquitous in the food safety regulatory world.

A visit to the *Codex Alimentarius* website will demonstrate to the reader the availability of hundreds of documents published by the *Codex*

---

2. Proposed Rule: Requirements for Additional Traceability Records for Certain Foods, 85 Fed. Reg. 59,984, 59,991 (Sept. 23, 2020).

Alimentarius Commission over the past half-century.[3] Perhaps the most important of all these documents ever published from a food safety perspective is the 1969 introduction of Hazard Analysis Critical Control Point (HACCP) methods to the food safety community. These HACCP principles are the foundation for the system of process control adopted virtually worldwide and the basis of the evolution of RBPCs—the intellectual basis for the FSMA in the United States, the Safe Food for Canadians Act, and similar regulations in the European Union. The reader is encouraged to study the *Codex* website in detail although the reader should understand that considerable background in food chemistry and toxicology will be necessary to fully understand *Codex* standards.

Additionally, disputes adjudicated under the dispute settlement provisions built into the WTO's governing treaties frequently become the governing issues in national food labeling regulatory systems. The WTO's website offers an excellent resource guide to the food-related aspects of these dispute resolution mechanisms and treaties.[4]

The "Work" of a U.S. "International Food Lawyer"

It is useful to describe the nature of a food lawyer's work in the international sphere as an illustration of the areas of competency that a client would expect from the attorney.

A finished food product comprises multiple component parts. For example, a cheeseburger, at minimum, has cheese, the protein, and a bun. But a food manufacturer must be keenly aware of where and how each of these items is sourced in order to provide clear and accurate information to regulators and consumers. Much of this reliance takes the form of compliance documentation used throughout the food chain.[5]

The following table provides an overview of some of the players involved in the logistical chain under FSMA from the perspective of an imported food to the United States.

---

3. To view all *Codex* standards, *see generally* Codex Alimentarius, *Standards*, http://www.fao.org/fao-who-codexalimentarius/codex-texts/list-standards/en/ (last visited May 4, 2021).

4. WTO & Organisation for Economic Co-operation and Development, Facilitating Trade through Regulatory Cooperation: The case of the WTO's TBT/SPS Agreements and Committees (2019), https://www.wto.org/english/res_e/booksp_e/tbtsps19_e.pdf.

5. Charles F. Woodhouse, *Food Lawyers Face Challenges from 21st Century Logistics, FSMA, and the Clean Label Movement*, A.B.A. Food Nutraceuticals & Cosmetics, Winter 2017, at 2.

Player	Role
Supplier	• Supply ingredients, food additives, and food contact materials • Provide technical specifications regarding same to downstream receivers
Overseas Manufacturer	• Manufacturing food product to ensure food safety—documenting the RBPC plan and Sanitary Transportation Plan
Agent (Customs Broker)	• Receive *documentation* from overseas manufacturer
First-Receiver in the United States	• Receives *goods* from overseas manufacturer—ultimate responsibility for FSMA rules compliance and traceability
U.S. Grocer or Food Service Distributor	• Sells food to end-user—maintains traceability records

At the start of the logistical chain, suppliers of ingredients, food additives, and food contact materials (i.e., food packaging) supply these component parts. In the case of products manufactured outside the United States, these overseas manufacturers must (1) maintain an RBPC plan, (2) have a sanitary transportation plan, (3) arrange its supply chain guarantees, and (4) comply with FDA's Foreign Supplier Verification Program, one of the many rules implemented pursuant to FSMA. Goods are transported to a United States-based importer who has ultimate responsibility for documentation of the transaction. That importer must also have available to FDA or USDA the RBPC plan and Sanitary Transportation Plan. Eventually, the goods head to the domestic U.S. customer, who often has indemnification rights pursuant to the underlying contract with the importer.[6]

These illustrative examples are from the perspective of the U.S. importer but are applicable to the obligations of the exporter as well, as we move from traditional HACCP principles into the modern era of RBPCs that are the foundations of new regulatory regimes in the United States, Canada, the European Union, China, and other modern countries.[7]

---

6. *See* Charles F. Woodhouse, *FSMA Final Rules on Sanitary Transportation of Human and Animal Foods*, A.B.A. FOOD NUTRACEUTICALS & COSMETICS, Spring 2016, at p. 4.

7. For an explanation of RBPCs see the continuing legal education (CLE) course materials that accompany the Lorman Education food law course presented each year by Charles F. Woodhouse. Lorman Education Services, https://www.lorman.com (last visited May 4, 2021).

Additionally, the U.S. attorney must make certain that clients are aware of the requirements of the FDA Reportable Food Registry.[8] There are significant civil and possibly criminal penalties applicable to individual executives in the supply chain for failure to timely report incidents as well as risks that product liability insurance coverage may be compromised.

How to Train to Become an "International" Food Attorney

In the table below, we present an illustration of the topics of the university course programs that are part of the necessary preparation of practitioners for the "international" aspect of food law.

Coursework
• Semester-length courses on food law and regulation in the United States, Canada, European Union, and China • FSMA compliance courses on preventive controls, the Produce Safety Rule, and, especially, the Foreign Supplier Verification Program • Coursework in microbiology, food toxicology, sanitization, and epidemiology—necessary if the practitioner is to prepare or professionally review RBPC plans

For my two graduate degrees in food safety and food packaging, some of the courses I have found most essential for my work as an "international food lawyer" have been on the topics of U.S. food regulation, Canadian food regulation, European Union food regulation, Chinese food regulation, and Latin American food regulation. For me, the best way for the reader to start acquiring this knowledge is through formal university training, as documented by earning graduate degrees from a research university.

This is the essential requirement for education—beyond a law degree—if you want to be a "food lawyer"—you will get there primarily through formal academic programs. The reader can take many of these courses listed at Michigan State, Johns Hopkins, University of California, Davis, or other research universities and can earn an M.S.-level graduate degree in two or three years (depending upon undergraduate science background necessary for admission to graduate-level science programs).

---

8. FDA, *Reportable Food Registry for Industry*, https://www.fda.gov/food/compliance-enforcement-food/reportable-food-registry-industry (last updated Sept. 10, 2020).

For those interested in diving into food law resources, I suggest the resources outlined in the following tables:

Membership Organizations	Journals
• Institute of Food Technologists (IFT) • International Association for Food Protection (IAFP) • American Agricultural Law Association (AALA) • Food and Drug Law Institute (FDLI)	• *Food Technology* (monthly) • *Food Safety Magazine* (bimonthly) • *FDLI Update* (quarterly)

News Sources	Trade Journals
• Food Safety News • Food Poisoning Bulletin • BarfBlog.com • Law blogs sponsored by various legal and academic organizations	• *Food Safety Magazine* • *The Packer*

Government Resources	International Agencies
• USDA Food Safety and Inspection Service • FDA warning letters and import alerts • Food Standards Agency (UK) • Health Canada • Canadian Food Inspection Agency	• *Codex Alimentarius* Commission • FAO • WHO

Focus Areas

A number of focus areas are central to my practice of international food law.

Discussion of Packaging and Food Contact Materials

There can be substantial food safety risks of current well-meaning but somewhat misdirected efforts to reduce packaging waste. This has resulted in a proliferation of attempts to introduce packaging waste restrictions in a variety of countries including certain municipalities in the United States, the European Union,[9] and Canada,[10] and has raised significant concerns in the food safety community.[11]

---

9. Press Release, European Parliament, Parliament Seals Ban on Throwaway Plastics by 2021 (Mar. 27, 2019), http://www.europarl.europa.eu/news/en/press-room/20190321IPR32111/parliament-seals-ban-on-throwaway-plastics-by-2021.

10. CANADIAN COUNCIL OF MINISTERS OF THE ENVIRONMENT, STRATEGY ON ZERO PLASTIC WASTE (2018) (PN 1583).

11. CHARLES F. WOODHOUSE, FOOD SAFETY ASPECTS OF THE EUROPEAN UNION SINGLE-USE PLASTICS BAN (2019) (essay in publication and in partial fulfillment of Michigan State University School of Packaging M.S. requirements).

The issues, of course, are both the proliferation of plastic packaging waste and "migration" into food products of chemical poisons. A discussion of the technical factors in food protection and packaging "shelf life," adsorption, permeation, and the selection of packaging polymers is beyond the scope of this chapter, as they require advanced knowledge of polymer engineering and macromolecular materials science. This knowledge can be best obtained by graduate study in materials science and engineering at institutions such as Michigan State University's School of Packaging, of which the author is an alumnus.

Since we do not wish to dismiss serious environmental and public health concerns (beyond the scope of this chapter), we refer the interested reader to an important article in *Science Advances* for detailed background.[12] We also urge the interested reader to familiarize himself or herself with the European Union's Circular Economy Action Plan.[13] As the United States rejoins the rest of the world's advanced economies, after a brief diversion into science denialism in public policy, we will hear much more about the circular economy initiatives that are now the norm among our major trading partners.

Even if the United States continues to resist circular economy planning and packaging materials reforms in other advanced countries, the $500 billion international food contact materials market will force U.S. companies (particularly those that sell in Canada and the European Union) to adopt many of the complex changes that are coming in the forms and uses of food packaging, including international standardization of polymer resins and plasticizers under sound environmental principles. We note that there are thousands of polymer extruders worldwide, but only a small number of resin manufactures that could be readily subject to international standardization norms.

Allergen Labeling

More than half of all U.S. food recalls are for label errors and more than half of those are for allergen label mistakes. This area is a major trap for the uninformed attorney and, again, we recommend formal academic training before advising clients. Failure to properly document allergens on food labels is a major problem triggering recalls in all major countries because of widely differing regulation, often associated with cultural

---

12. Roland Geyer et al., *Production, Use, and Fate of All Plastics Ever Made*, 2017 SCIENCE ADVANCES e1700782 (2017).

13. European Commission, *First Circular Economy Action Plan*, https:// ec.europa.eu/environment/circular-economy/ (last visited May 4, 2021).

factors in the diets of consumers. Lack of professional knowledge of allergens is a major malpractice trap for attorneys.

At the time of this writing, President Biden has recently signed the Food Allergy Safety, Treatment, Education, and Research (FASTER) Act of 2021, which adds sesame as the ninth major allergen designated by FDA.[14]

For example, FDA lists 9 major allergens, Canada lists 12, and the European Union lists 14 major allergens for mandatory label disclosure to consumers, as outlined in the table below.

Major Allergens in the United States	Major Allergens in Canada	Major Allergens in the European Union
• Milk • Eggs • Fish • Shellfish • Tree nuts • Peanuts • Wheat • Soybeans • Sesame	• Eggs • Milk • Mustard • Peanuts • Crustaceans and molluscs • Fish • Sesame seeds • Soy • Sulfites • Tree nuts (almonds, Brazil nuts, cashews, hazelnuts, macadamia nuts, pecans, pine nuts, pistachios, and walnuts) • Wheat and triticale • Gluten	• Cereals containing gluten • Crustaceans and products thereof • Eggs • Fish • Peanuts • Soybeans • Milk • Nuts • Celery • Mustard • Sesame seeds • Sulfur dioxide (SO_2) and sulfites at concentrations of more than 10 mg/kg or 10 mg/liter in terms of the total SO_2 • Lupin • Molluscs

The Food Allergy Research and Resource Program, at the University of Nebraska-Lincoln, maintains a regularly updated international food allergen spreadsheet covering more than 25 countries.[15] This chart should be on the desk or iPad of all food law practitioners.

---

14. FASTER Act, Pub. L. No. 117-11, 135 Stat. 262 (2021).

15. University of Nebraska-Lincoln, *Food Allergens—International Regulatory Chart*, https://farrp.unl.edu/IRChart (last updated Aug. 21, 2020).

Pesticide and Biocide Residual Levels

Researching and determining country-specific maximum residual levels for natural and anthropogenic poisons in major export markets is a problem for which the legal practitioner must have advanced science credentials that are obtained only through graduate-level education. The nonscientist lawyer must be cautious, as this is a major source of customs, health regulatory, and end-user rejections. Finding and accessing this information—indeed even knowing what to look for—is greatly facilitated by being an enrolled graduate student at a major research institution. Michigan State University has built detailed databases on these topics that are accessible to enrolled graduate students.

Food Labeling—International Trade

Unfortunately, the international food law practitioner can no longer rely upon what has been regarded as definitive information and guidance on general labeling issues in the FDA Food Labeling Guide.[16] This document does not incorporate the most significant labeling changes mandated by FSMA over the past eight years. Thus, U.S. practitioners are sent back to the complexities of 21 C.F.R. part 101 (Food Labeling). Labeling reviews can be highly detailed and fact-specific and present major professional liability risks to lawyers who have not pursued extensive training.

Food Safety Documentation

Each of your food industry clients who exports or imports animal or human foods or food ingredients must understand the transition of regulatory authorities in the United States, Canada, the European Union, and most modern countries away from HACCP and toward the new era of RBPCs.[17] The author presents a continuing legal education (CLE) course on FSMA and food law through Lorman Education and students are given extensive materials that discuss the RBPC concept in great detail.

Recalls and Recall Management

A central portion of the tasks of an international food lawyer is client counseling on dealing with product recall situations in specific jurisdictions. While individual seizures and testing delays related to border

---

16. FDA, A Food Labeling Guide: Guidance for Industry (2013).
17. Current Good Manufacturing Practice, Hazard Analysis, and Risk-Based Preventive Controls for Human Food, 80 Fed. Reg. 55,908 (Sept. 17, 2015). For a general discussion of the FSMA RBPCs see Charles F. Woodhouse, *Preparing for the Food Safety Modernization Act*, A.B.A. SciTech Law., Summer 2014, at p. 24.

inspections are a recurring problem, product recalls can present a major existential threat to the food business. Swiss Reinsurance studies have placed the annual cost of food recalls for U.S. business at more than $15 billion annually.[18]

Government resources are often the best source for food recall activity. In the United States, USDA and FDA publish details of major food recalls. In the European Union, annual reports of the Rapid Alert System for Food and Feeds (RASFF) are a great resource. In Canada, the Canadian Food Inspection Agency posts food recall warnings and allergy alerts. In addition to government resources, Stericycle publishes a detailed analysis of U.S. food recalls on a quarterly basis, with each issue having a separate chapter on food and beverage recalls.[19]

Without regularly following regulatory actions in the countries of interest to one's clients, the attorney will not be able to adequately fulfill the "prediction" role so well articulated by Justice Holmes. An international food lawyer should be able to tell his or her client not just "where we are," but also "where we are going."

Specific Issues in European Union Food Law Research

As a note to future international food law practitioners, the regulation of food in the European Union is an extremely complex and directed course of study, and a dedicated course on European food law and regulation is strongly suggested. Of course, the place to start would be the foundational legislation that underlies the structure of European Union food regulation—Regulation (EU) No. 1169/2011, on the provision of food information to consumers,[20] and Regulation (EC) No. 852/2004, on the hygiene of foodstuffs.[21] For those without formal university training in European Union law, we recommend starting with the excellent orientation provided by Fordham Law School's Citation Manual for European Union Materials.[22] Also useful for accessing European Commission and

---

18. Swiss Re, *Food Safety in a Globalized World, 2019*, https://corporatesolutions.swissre.com/insights/knowledge/food-safety-in-a-globalised-world.html.

19. Stericycle, *Recall Index 2020*, https://marketing.sedgwick.com/acton/media/4952/us-q3-2020-recall-index.

20. Council Regulation 1169/2011 on the Provision of Food Information to Consumers, 2011 O.J. (L 304) 18.

21. Council Regulation 852/2004 on the Hygiene of Foodstuffs, 2004 O.J. (L 139) 1.

22. *A Citation Manual for European Union Materials: 2010–2011 Edition*, 34 FORDHAM INT'L L.J. 1 (2010/2011).

European Parliament materials is the European Union's Interinstitutional Style Guide.[23]

Conclusion

A major challenge for the international food lawyer is to guide clients as to where we are going with active, informative, and intelligent packaging and preservation systems in a world that is moving toward a sustainable, circular economy-compliant, international food system. The most significant developments in food regulation are coming from our trading partners and will require sophisticated understanding and response from U.S. stakeholders if they hope to prosper in a worldwide food commerce system.

Get educated, stay informed, know the essential information sources, never stop learning and studying, and build your professional network of contacts.

---

23. EUROPEAN UNION, INTERINSTITUTIONAL STYLE GUIDE (2011).

Chapter 8

Blockchain in the Food Industry

Darin Detwiler

Introduction

Today's stakeholders across the entire spectrum of seed to fork live in a world where technology is all around them. Those in the industry may see it in the form of software, robotics, satellite imagery, and even virtual reality. Consumers may see advances in food technology in the way they gain information about food, share feedback, order food, and even in how they pay for food.

Over time, systems of food production have changed dramatically. From ancient eras to modern times, one thing is certain: people need food to eat. In the arc of history, industrialization has changed the process of food production and technology has shaped how consumers obtain their food. Technology has also influenced how regulators oversee the food system. The needs of food safety systems have also changed over time. An increasingly global marketplace poses significant challenges for producers and for regulators in ensuring the safety of our food supply.

Officials from the U.S. Food and Drug Administration (FDA) recently addressed how the agency was adapting its expectations for advancements in technology.[1] Writing in 2019, FDA leadership announced that

---

1. Press Announcement, FDA, Statement from Acting FDA Commissioner Ned Sharpless, M.D., and Deputy Commissioner Frank Yiannas on Steps to Usher the U.S. into a New Era of Smarter Food Safety (Apr. 30, 2019), https://www.fda.gov/news-events/press-announcements/statement-acting-fda-commissioner-ned-sharpless-md-and-deputy-commissioner-frank-yiannas-steps-usher.

the agency expected "see more innovation in the agriculture, food production, and food distribution systems in the next 10 years than [it has] seen in the past 20, which will continue to provide an even greater variety of food options and delivery conveniences to American consumers."[2] Given this "ever-changing landscape," agency leadership counseled that FDA "must continue preparing to take advantage of new opportunities and address potential risks."[3]

A "New Era of Smarter Food Safety"

To meet increasing food safety needs with advancements in technology, FDA announced what the agency is calling a "New Era of Smarter Food Safety." The agency has released its blueprint to this New Era of Smarter Food Safety with the announcement that the nation "is in the midst of a food revolution," requiring modern approaches for modern times.[4] Then-FDA Commissioner Stephen Hahn noted: "Smarter Food Safety to me means always looking to the future. Our destination—safe food for our families, our children, and our animals—is unchanged. But how do we get there more quickly and effectively using modern tools as the world transforms around us?"[5] Amid changing technologies and new challenges, the agency has made its goal clear: "FDA and [its] stakeholders should be looking at how to tap into new technologies that include, but are not limited to, artificial intelligence, the Internet of Things, sensor technologies, and blockchain."[6]

According to the agency, this "New Era" is centered on four core elements:

Tech-Enabled Traceability. This would assist responses to food safety incidents, especially outbreaks. FDA noted that challenges to properly identify, track, and trace food during food safety outbreaks can "cost lives, millions of dollars in avoidable product loss, and damage to consumer trust."[7] Technology can create a "more nimble, resilient, and interoperable food system."[8]

Smarter Tools and Approaches for Prevention and Outbreak Response. The agency is looking to technology to improve food

---

2. *Id.*
3. *Id.*
4. FDA, New Era of Smarter Food Safety: FDA's Blueprint for the Future (2020), *available at* https://www.fda.gov/media/139868/download.
5. *Id.* at 2.
6. *Id.*
7. *Id.*
8. *Id.*

traceback, as well as its ability to conduct root-cause analyses to create a preventive controls system that incorporates new information.[9] The agency is also aiming to make "processes and communications more effective, efficient, and in some cases, simpler."

New Business Models and Retail Modernization. FDA is examining how changes to food production and distribution are changing business models, such as increasing amounts of online grocery shopping. The agency is examining how it can best educate distributors, manufacturers, and retailers on the importance of temperature control, cross-contamination, and other safety issues, as well as how it can adapt its oversight to help ensure the safety of novel ingredients, new foods, and new food production methods.[10]

Food Safety Culture. The agency is looking at how it can foster, support, and strengthen food safety culture on farms, in food facilities, and in homes, as it acknowledges that it will "not make dramatic improvements in reducing the burden of foodborne disease without doing more to influence and change human behavior, addressing how employees think about food safety and how they demonstrate a commitment to this goal in how they do their job."[11] The COVID-19 pandemic further emphasized to the agency the importance of a food safety culture given the agency's focus on keeping food workers safe and educating consumers cooking more at home on safe food handling practices.

Food Safety History

Emerging technologies like blockchain have drastically changed food systems across all of human history. Food technology in its earliest forms can be found in the use of fire, circa 700,000 B.C., and fermentation, circa 10,000 B.C. Food technologies helped prompt societal shifts from nomadic living to the domestication of animals and early agriculture.

The following table provides a summary of how food technologies have changed over time.[12]

---

9. *Id.*
10. *Id.*
11. *Id.*
12. Based on information from KLAUS SCHWAB, THE FOURTH INDUSTRIAL REVOLUTION (2017); Michael Haupt, *Society 4.0: The Evolutionary Journey to Humanity's Next Transition*, MEDIUM, Jan. 21, 2018, https://medium.com/society4/evolution-of-societies-93a5f0f9b31; Michael Siegrist & Christina Hartmann, *Consumer Acceptance of Novel Food Technologies*, 1 NATURE FOOD 343 (2020), *available at* https://doi.org/10.1038/s43016-020-0094-x.

Era	Approximate Dates	Selected Food Technologies
Pre-Industrial Revolution	Prior to 1700s	Fire (700,000 B.C.) Fermentation (10,000 B.C.)
First Industrial Revolution	18th to 19th centuries	Canning (1790s) Pasteurization (1830s) Refrigeration (1850s–1900)
Second Industrial Revolution	1870–1917	Industrialization/centralization Regulation of food manufacturing
Post-Second Industrial Revolution	1917–1980s	Frozen foods/fast foods Microwaving foods
Third Industrial Revolution	1980s–today	Computers/Internet technologies Genetically modified organisms/biotechnology Precision agriculture
Fourth Industrial Revolution	2010s and beyond	Robotics/artificial intelligence Nanotechnology/quantum computing Internet of Things Blockchain

Even in 1906, society recognized that food safety issues are endemic, occurring in the past, at the present, and would occur in the future. The *London Times* wrote, in discussing Upton Sinclair's *The Jungle*: "The things described by Mr. Sinclair *happened yesterday, are happening today, and will happen tomorrow and the next day*, until some Hercules comes to cleanse the filthy stable."[13]

Technology has always been part of how humans consume and produce food. Society has advanced from hunter-gatherers to systems of agriculture, from hand tools to steam engines, from analog to digital platforms, and even from horse-drawn carriages to unmanned drones for delivery. The changes in food technologies have prompted significant transformations in the geography of how food is produced (e.g., distances for import/export) and consumed (e.g., eating outside the home and, now, ordering delivery over the smartphone). They have impacted the economics of farmers, producers, and consumers in myriad ways beyond simple calculations of economic efficiency. They have also resulted in the demand for legal and policy changes to promote public health. Further, these technological changes to the food system have transformed the relationship between individual consumers and the food they eat.

---

13. *The Jungle*, TIMES LITERARY SUPPLEMENT (1906) (microfilm collection, Western Washington University) (emphasis added).

The "Five Pillars" of the Food System

Food safety is a critical piece of the food system, but it is not the only component. Together with (1) food safety, (2) food quality, (3) food authenticity, (4) food defense, and (5) food security constitute the five pillars of the overall food system.[14]

Put another way, technology assists stakeholders in the food system in meeting these five pillars to keep food (1) safe to eat, (2) at an appetizing quality, (3) free from fraud, (4) defended against threats that compromise human health, and (5) available and accessible for all. These five pillars are summarized in the table below.[15]

The Five Pillars of the Food System	
Food Safety	• Physical contamination • Chemical contamination • Biological contamination
Food Quality	• Taste • Smell • Appearance
Food Fraud (Authenticity)	• Adulteration • Tampering • Diversion • Simulation • Counterfeiting
Food Defense	• Industrial sabotage • Terrorism • Economically motivated adulteration
Food Security	• Availability • Utilization • Access • Stability

Food Safety

Food safety efforts play a vital role in detection of hazards and, thus, in preventing adverse impacts on the health of humans or animals. Proactive steps are paramount in protecting consumers.

Pursuant to guiding industry standards, "food safety" is commonly considered to include:

---

14. John G. Keogh et al., *Optimizing Global Food Supply Chains: The Case for Blockchain and GS1 Standards*, in BUILDING THE FUTURE OF FOOD SAFETY TECHNOLOGY: BLOCKCHAIN AND BEYOND 171 (Darin Detwiler ed., Academic Press 2020), *available at* https://www.sciencedirect.com/science/article/pii/B9780128189566000178.

15. *Id.*

Food Safety System Certification (FSSC): FSSC-0-006.2:

Food/Feed Safety—The policies, processes and procedures, materials, facilities and monitoring systems applied to Food or Feed products to ensure they will not cause harm to humans or animals or adversely affect their health when utilized according to their intended purpose.[16]

FSSC-0-006.3:

Food/Feed Safety Hazard—Biological, chemical, physical agent or allergen in food/feed, or condition of food/feed, with the potential to cause an adverse health effect to humans and/or animals (ISO 22000, section 3.3).[17]

Food Quality

Measures of food quality date back to the earliest days of bartering. Organoleptic testing (use of sensory organs for evaluations of the odor, flavor, and texture of food) served for nearly a century as the foundation of regulators' work in detecting foodborne pathogens before the advancement of more modern methods, such as genomics.

Common to most definitions is that food quality involves processes to support consistent manufacturing specifications and desired product characteristics, including sensory (e.g., density, color, smell, taste, texture, viscosity, etc.) and environmental qualities (e.g., place of sale, packaging, storage, etc.).

Food Authenticity

Food authenticity—ensuring food is not misbranded or adulterated—demands vigilant actions regarding auditing and validation. Public health can be adversely affected long before any regulator or court will determine whether an act is intentional, economically motivated, or worse. FDA leadership stated at a 2016 food industry conference that the agency will step in when food authenticity issues threaten public health.[18]

---

16. Darin Detwiler, *Food Fraud and Food Defense*, in FOOD SYSTEM TRANSPARENCY: LAW, SCIENCE, AND POLICY OF FOOD AND AGRICULTURE 19 (Gabriela Steier & Adam Friedlander eds., CRC Press 2021).

17. *Id.*

18. Stephen Ostroff, FDA's Take on Criminal Liability, Presentation at Plenary Session and Town Meeting Conducted at the Food Safety Consortium (Nov. 2016).

Food authenticity can be considered from a variety of angles. For example, the Food Safety Modernization Act Final Rule for Mitigation Strategies to Protect Food against Intentional Adulteration noted:

> Mitigation Strategies to Protect Food Against Intentional Adulteration are *"aimed at preventing intentional adulteration from acts intended to cause wide-scale harm to public health, including acts of terrorism targeting the food supply. Such acts, while not likely to occur, could cause illness, death, [and] economic disruption of the food supply absent mitigation strategies."*[19]

Food authenticity may be considered in terms of seven different risks:[20]

- **Adulteration.** Mixing matter of an inferior and sometimes harmful quality with food intended to be sold. As a result, it becomes impure and unfit for human consumption (aka dilution).
- **Tampering.** Legitimate product and packaging are used in a fraudulent way (aka misbranding).
- **Overrun.** Legitimate product is made in excess of production agreements.
- **Theft.** Legitimate product is stolen and passed off as legitimately procured.
- **Diversion.** The sale or distribution of legitimate products occurs outside of intended markets.
- **Simulation.** One product is designed to look like the real, labeled product (aka substitution).
- **Counterfeiting.** This entails intellectual property rights infringement, including fraudulent product or packaging.

Food Defense

Food defense—defending the food supply against economically motivated external threats—is a critical part of securing the food system. Unfortunately, public health issues may occur long before any intentional act against food is identified. Regular, sustained actions to defend the food supply from external threats serves as a deterrent, but unfortunately issues do occur that present real and present dangers for consumers' health. For example, a Russian counterfeiting ring was tied to a

---

19. NATIONAL CENTER FOR FOOD PROTECTION AND DEFENSE, BACKGROUNDER: DEFINING THE PUBLIC HEALTH THREAT OF FOOD FRAUD (2011), https://learn.canvas.net/courses/1674/files/623752/download?download_frd=1.
20. *Id.*

fake infant formula incident in New Zealand in 2016.[21] In 2008, the use of melamine in baby food from China resulted in 300,000 victims and 54,000 babies hospitalized.[22]

Definitions of "food defense" include:
FSSC 22000-0-005.2:

Processes to prevent—"food and feed supply chains from all forms of ideologically or behaviorally motivated, intentional adulteration that might impact consumer health."[23]

FSSC-2-006.1: Food Defense 2.1.4.5.2:

Preventive measures . . . The organization shall put in place appropriate preventive measures to protect consumer health impacts.[24]

Food defense can be defined in terms of three different risks:[25]

- **Industrial sabotage.** This is the intentional contamination by an insider or competitor to damage the company, causing financial problems/a recall but not necessarily to cause public harm.
- **Terrorism.** The reach and complexity of the food system has caused concern for its potential as a terrorist target.
- **Economically motivated adulteration.** This includes acts against a product for the purpose of increasing the apparent value of the product or reducing the cost of its production (i.e., for economic gain).

Food Security

Food security is an issue of major international concern. The United Nations General Assembly has set 17 Sustainable Development Goals

---

21. Matthew Theunissen, *New Kiwi Software Intercepts Fake Infant Formula*, N.Z. HERALD, Dec. 17, 2016, http://www.nzherald.co.nz/business/news/article.cfm?c_id=3&objectid=11769310.

22. *See* Yanzhong Huang, *The 2008 Milk Scandal Revisited*, FORBES, July 16, 2014, http://www.forbes.com/sites/yanzhonghuang/2014/07/16/the-2008-milk-scandal-revisited/#41a6bb204428; Jane Macartney, *China Baby Milk Scandal Spreads as Sick Toll Rises to 13,000*, TIMES, Sept. 22, 2008, https://www.thetimes.co.uk/article/china-baby-milk-scandal-spreads-as-sick-toll-rises-to-13000-jlxdmrsk9qd.

23. JOHN SPINK, MICHIGAN STATE UNIVERSITY FOOD FRAUD INITIATIVE, FFI REPORT: REVIEW—NEW FSCC 22000 VERSION 4 REGARDING FOOD FRAUD AND FOOD DEFENSE (2016), https://www.blog.foodfraudpreventionthinktank.com/wp-content/uploads/2021/02/MSU-FFTT-FFIR-FSSC-22000-update-Edition-4-2017-v8.pdf.

24. *Id.*

25. NATIONAL CENTER FOR FOOD PROTECTION AND DEFENSE, *supra* note 19.

(SDGs) for the year 2030. Though all the 17 goals have a connection to food,[26] some SDGs have names that provide a clearer connection to food, such as Zero Hunger, Good Health and Well-Being, Clean Water and Sanitation, Sustainable Cities and Communities, Responsible Consumption and Production, Climate Action, Life below Water, and Life on Land. These SDGs reflect the challenges that technological advances in the Fourth Industrial Revolution are poised to address.

Food security aims to afford individuals access and availability to safe and nutritious food. For example, the United Nations' Food and Agriculture Organization declared, "Food security exists when all people, at all times, have physical, social and economic access to sufficient, safe and nutritious food which meets their dietary needs and food preferences for an active and healthy life."[27]

Food security can be defined in terms of four different risks:[28]

- **Availability**. This includes the supply of food through production, distribution, and exchange.
- **Access**. Access describes the affordability and allocation of food, as well as the preferences of individuals and households.
- **Utilization**. The metabolism of food by individuals can be affected by food safety, nutritional values, food choice, and cultural preferences.[29]
- **Stability**. This describes the ability to obtain food over time, as food insecurity can be transitory, seasonal, or chronic. It can be impacted by failures in food sustainability, defense, and so on.

Blockchain and the Greater Technology Ecosystem

The Food Technology Ecosystem

As part of the Fourth Industrial Revolution, discussions are evolving our understanding of how technology can support the five pillars of the food

---

26. *How Food Connects All the SDGs*, STOCKHOLM RESILIENCE CENTRE, June 14, 2016, https://www.stockholmresilience.org/research/research-news/2016-06-14-how-food-connects-all-the-sdgs.html.

27. UNITED NATIONS FOOD AND AGRICULTURE ORGANIZATION, COMMODITY POLICY AND PROJECTIONS SERVICE: COMMODITIES AND TRADE DIVISION § 2.2 (2003), *available at* http://www.fao.org/3/y4671e/y4671e06.htm.

28. *How Food Connects All the SDGs*, *supra* note 26.

29. Note that this "utilization" component of food security depends on validation and verification for food safety, nutritional values, food choice, and cultural preferences. Foods pertaining to diets—such as kosher, halal, vegan, vegetarian, allergy conscious, impacted immune status, etc.—require food compliance validation from retail and their suppliers.

system discussed above. New technologies have prompted changes to consumer expectations, as consumers "can now expect to see the entire histories of the products they buy, and hence make more informed decisions."[30]

For example, barcodes and QR codes are often embedded with information for use by the retailer and the consumer. Increasingly, retailers are using barcodes not only to more efficiently process consumers' purchases at the checkout line but also to manage inventory and promote traceability throughout the supply chain.

From the launch of barcodes in the 1970s, federal and international regulators only had a paper-based requirement for proof of production, sales, and distribution.[31] Barcodes and now QR codes have evolved to relay more information from the farm to retailers and to consumers. However, "regulators still remained static in their requirement and suggestion of a paper standard for one-up-one down traceability and accountability"[32] For years, digitization of these records was merely a *suggestion*.

The reliance on paper records can complicate the investigation of food safety outbreaks. Then-FDA Commissioner Scott Gottlieb noted that paper records made the large-scale investigation of multiple outbreaks tied to romaine lettuce and other leafy green vegetables in 2018 more difficult.[33]

Paper records have been the industry standard, but times are changing. The past decade has seen a sea change as industry and government stakeholders transition to more technically advanced methods of record-keeping to promote tracking and traceability throughout complex food supply chains. For example, Walmart announced in 2018 that it would require real-time, end-to-end food traceability from their suppliers of leafy green vegetables.[34]

---

30. Saif Rivers, *Win Customers' Hearts with a Transparent Supply Chain*, BLOCKCHAIN PULSE: IBM BLOCKCHAIN BLOG, Apr. 22, 2019, https://www.ibm.com/blogs/blockchain/2019/04/win-customers-hearts-with-a-transparent-supply-chain/.

31. Darin Detwiler, *The Evolution of Food Safety: From Barcode to Blockchain*, NE. U. GRADUATE PROGRAMS BLOG, Nov. 28, 2019, https://www.northeastern.edu/graduate/blog/food-safety-evolution/.

32. *Id.*

33. Press Announcement, FDA, Statement from FDA Commissioner Scott Gottlieb, M.D., on Findings from the Romaine Lettuce E. coli O157:H7 Outbreak Investigation and FDA's Efforts to Prevent Future Outbreaks (Nov. 1, 2018), https://www.fda.gov/news-events/press-announcements/statement-fda-commissioner-scott-gottlieb-md-findings-romaine-lettuce-e-coli-o157h7-outbreak.

34. Press Release, Walmart, Walmart and Sam's Club to Require Real-Time, End-to-End Food Traceability with Blockchain (Sept. 24, 2018), https://corporate

Now, with the "New Era of Smarter Food Safety," FDA acknowledged that all stakeholders in the food supply chain must "continue preparing to take advantage of new opportunities and address potential risks" brought about by continuing innovation in agriculture, food production, and food distribution systems that are poised to "provide an even greater variety of food options and delivery conveniences to American consumers."[35] This ever-changing landscape demands the use of "new and emerging technologies" to "address several areas, including traceability, digital technologies and evolving food business models."[36] FDA now has essentially aligned policy with new technologies, such as blockchain.

Blockchain

Blockchain offers an immutable, decentralized eLedger: a collection of blocks of information for each step, farm to fork.[37] This digital information lives in multiple databases at once, making it extremely difficult to edit, change, or forge. Being immutable, blockchain creates a permanent record, which can be referenced quickly and with greater confidence than traditional records in the event of an emergency.[38]

Food companies are already using blockchain to promote traceability in their supply chains. For example, Walmart announced:[39]

> Blockchain is a way to digitize data and share information in a complex network in [a] secure and trusted way. For food safety, this helps to more accurately pinpoint issues in the food chain and further protect customers against foodborne illnesses.
>
> ...
>
> For more than a year, Walmart, working with IBM and 11 other food companies, has successfully developed a blockchain-enabled food traceability network built on open-source technology. In an initial pilot conducted by Walmart and IBM, the amount of time it took the retailer to trace an item from store to farm was reduced from seven days to just 2.2 seconds.

.walmart.com/media-library/document/leafy-greens-on-blockchain-press-release/_proxyDocument?id=00000166-0c4c-d96e-a3ff-8f7c09b50001.

35. Press Announcement, FDA, *supra* note 1.
36. *Id.*
37. Darin Detwiler, *Is Blockchain the Traceability Solution?*, QUALITY ASSURANCE & FOOD SAFETY, Aug. 7, 2018, https://www.qualityassurancemag.com/article/is-blockchain-the-traceability-solution/.
38. *Id.*
39. Press Release, Walmart, *supra* note 34.

In the words of IBM: "Blockchain offers complete visibility of the data behind the many stages of product creation. Manufacturers, farmers, wholesalers, suppliers, delivery services and stores each input information that details and verifies their roles in the process—creating a log that provides irrefutable evidence of a product's provenance."[40]

Each transaction or set of transactions becomes a block that are arranged in a chain, a blockchain. From the first block—the "genesis" block recording the first transaction—new blocks are added containing new information and linking back to the previous block. Because duplicate copies of the database exist on multiple computers, these computers all need to agree on the blocks as they are recorded, making it more difficult if not impossible to modify, manipulate, or fake the transactions once they are recorded in a block.

Blockchain allows for the creation of smart contracts that promote traceability across the food supply chain and provide industry and regulatory officials the ability to trace food items back all the way to their source to assist in investigations and recall efforts. That said, the utility of blockchain in the food industry is not solely limited to preventing failures in food safety. Food quality, food authenticity, food defense, and food security all involve similar needs related to transparency and traceability supported by these same efforts through digitized records and analysis of data shared via blockchain.

A simplified model of how blockchain can be used to promote food traceability is provided in the following table.

---

40. Rivers, *supra* note 30.

Blockchain Access and Mapping—Simplified Example		
Stakeholder	**Additions to the Block**	**Traceback to:**
Food Growers/ Producers	Genesis block • Water testing • Soil testing • Health/safety documentation	
Food Manufacturers/ Processors	• Audits • Testing results • Compliance certifications	Food growers/producers
Distributors	• Audits • Environmental monitoring • Compliance certifications	Food growers/producers Food manufacturers/ processors
Retailers and Food Service	• Inspections • Environmental monitoring • Compliance certifications	All earlier entities in the block
Consumers	Limited information access	All earlier entities in the block

Put another way, let's say Frank's Food Store wants to purchase a product from Manny's Manufacturing and ships their product through Sam's Shipping.[41] First, the manufacturer and the store enter into a smart contract whereby the store puts money in escrow to be released upon delivery of the product ordered. Then the manufacturer and the shipping company could enter into their own smart contract whereby Manny's Manufacturing puts money into escrow for Sam's Shipping to pick up said product and deliver it to the Frank's Food Store. Beyond just moving the product from point A to point B, additional rules or conditions could include delivery times, temperature requirements, pathogen testing, and audit results. All of these agreements are programmed into the smart contract and stored on the blockchain.

---

41. This example is adapted from Gennette Zimmer, *Defining Terms*, in BUILDING THE FUTURE OF FOOD SAFETY TECHNOLOGY: BLOCKCHAIN AND BEYOND 3 (Darin Detwiler ed., Academic Press 2020), *available at* https://www.sciencedirect.com/science/article/pii/B9780128189566000014.

If Frank's Food Store receives the product, and all the rules in the smart contracts were adhered to, the escrowed funds are released to both Manny's Manufacturing and Sam's Shipping. If not, the product is not accepted and the escrowed funds are returned.

Blockchain technology and smart contracts are currently being explored in business operations more broadly. In February 2019, then-FDA Commissioner Scott Gottlieb announced that the agency will be testing new technologies to enhance industry track-and-trace systems for the *drug* supply chain. In his statement, Commissioner Gottlieb noted that the agency was "focused on making improvements across the other products we regulate, especially related to food and our ability to address foodborne outbreaks. We're invested in exploring new ways to improve traceability, in some cases using the same technologies that can enhance drug supply chain security, like the use of blockchain."[42]

FDA's "New Era of Smarter Food Safety" has recommended that stakeholders "implement an internal digital technology system, such as blockchain, to receive critical tracking events and key data elements from industry and regulatory partners."[43] Even with substantial momentum, some predict that widespread implementation of blockchain in food and beverage supply chains may still be decades away.[44]

Conclusion

As FDA has noted, modern times call for modern approaches to the food system. Indeed, the most significant food statute of the recent past was entitled the Food Safety *Modernization* Act. From *The Jungle* to adulterated baby food sickening tens of thousands just years ago, society must wrestle with the very real threats related to the food system. Unfortunately, history tells us that the most significant changes in food regulations and industry practices came about after the serious illnesses and deaths of young children.[45]

---

42. Press Announcement, FDA, FDA Takes New Steps to Adopt More Modern Technologies for Improving the Security of the Drug Supply Chain through Innovations That Improve Tracking and Tracing of Medicines (Feb. 7, 2019), https://www.fda.gov/news-events/press-announcements/fda-takes-new-steps-adopt-more-modern-technologies-improving-security-drug-supply-chain-through.

43. FDA, *supra* note 4.

44. Darin Detwiler, Food Safety: Past, Present, and Predictions (2020).

45. *Id.* at 238–39.

Technological advances, such as blockchain, hold significant promise to promote the food system's five pillars. While it may be several years before industry-wide implementation, efforts are already underway to utilize blockchain to empower all stakeholders in the food system, promote traceability of food products, and improve food business operations.

Chapter 9

Food Law and the Pandemic: Securing the Food System

Lawrence Reichman and Tommy Tobin

The COVID-19 pandemic has upended the food supply chain, how consumers shop, and how government agencies approach their oversight of the food system. From families to food manufacturers, the economic consequences of the pandemic are real, and many shifts may persist for years to come. This chapter outlines the role of federal food-related agencies in an all-of-government response to the novel coronavirus's effects on the American economy and the changes it has brought about in daily life for food producers and consumers.

COVID-19 from the Federal Agency Perspective

The response to the COVID-19 pandemic extended beyond any one agency. The consequences of the pandemic, both in economic cost and the cost to human life, have been nothing short of staggering. Not only did the pandemic drastically shift the operations of government regulators, but its economic effects have impacted both how consumers obtain and how producers supply food. As of March 2021, there were more than 30 million cases of COVID-19, the disease caused by the novel coronavirus, in the United States.¹ More than 500,000 Americans have died because of the disease.²

Many American families experienced layoffs or furloughs. While Congress has passed several pandemic-related relief bills,³ families around the country struggled to access affordable and healthful food.⁴ Leading researchers reported that the "COVID-19 crisis has already left too many children hungry in America" and "young children are experiencing food insecurity to an extent unprecedented in modern times."⁵ Not only did the COVID-19 pandemic's economic consequences lead to long lines for food assistance,⁶ they unfortunately deepened disparities across racial and ethnic lines.⁷ During the pandemic, Black and Latinx households experienced food insecurity at rates more than 50 percent higher than those of white households.⁸

The food system has faced immense issues during the pandemic. The nation's food system is highly fragmented. At the federal level, the U.S. Food and Drug Administration (FDA) oversees food safety for about 80 percent of the nation's food products. The U.S. Department of Agriculture (USDA) governs the remaining quintile, focusing on shell eggs, meat

---

1. Centers for Disease Control and Prevention, *COVID Data Tracker: United States COVID-19 Cases and Deaths by State*, https://www.cdc.gov/covid-data-tracker/index.html#cases (accurate as of Mar. 10, 2020).

2. *Id.*

3. *See* Families First Coronavirus Response Act, Pub. L. No. 116-127, 134 Stat. 178 (2020); Coronavirus Aid, Relief, and Economic Security Act (CARES Act), Pub. L. No. 116-136, 134 Stat. 281 (2020); American Rescue Plan Act of 2021, Pub. L. No. 117-2, 135 Stat. 4.

4. Tommy Tobin, *Fighting Food Insecurity amid COVID-19, Groups Advocate for 15% Food Stamp Boost*, FORBES, July 8, 2020, https://www.forbes.com/sites/tommytobin/2020/07/08/fighting-food-insecurity-amid-covid-19/#1f7224466862, *archived at* https://perma.cc/6DJV-UTG3.

5. Lauren Bauer, *The COVID-19 Crisis Has Already Left Too Many Children Hungry in America*, BROOKINGS: UP FRONT, May 6, 2020, https://www.brookings.edu/blog/up-front/2020/05/06/the-covid-19-crisis-has-already-left-too-many-children-hungry-in-america/, *archived at* https://perma.cc/KLE8-STHK.

6. Tracey Tully, *Food Lines a Mile Long in America's Second-Wealthiest State*, N.Y. TIMES, Apr. 30, 2020, https://www.nytimes.com/2020/04/30/nyregion/coronavirus-nj-hunger.html.

7. Helena Bottemiller Evich, *Stark Racial Disparities Emerge as Families Struggle to Get Enough Food*, POLITICO, July 6, 2020, https://www.politico.com/news/2020/07/06/racial-disparities-families-struggle-food-348810.

8. *Id.*

and poultry products, and catfish.[9] Other agencies are also involved in regulating certain aspects of the labeling of food products, such as the Federal Trade Commission.

FDA leadership highlighted the agency's "active leadership role in the *all-of-government* response to the COVID-19 pandemic," noting that FDA had created an internal cross-agency group to ensure the agency was "doing everything possible to protect the American public, help[ing] ensure the safety and quality of FDA-regulated products, and provid[ing] the industries [it] regulate[s] the tools and flexibility to do the same."[10] While noting its many efforts regarding the oversight of vaccine development and FDA-regulated drug products, the agency also exercised regulatory authority "to protect consumers from firms and individuals selling unproven products with false or misleading claims to prevent, treat, mitigate, diagnose, or cure COVID-19, including by issuing warning letters and pursuing civil and criminal enforcement actions, where appropriate."[11] FDA (1) sent complaints to online marketplaces and domain name registrars to voluntarily remove listings for products that fraudulently claim to diagnose, cure, mitigate, treat, or prevent COVID-19; (2) issued more than 110 warning letters to food, drug, and dietary supplement companies selling such fraudulent products; and (3) referred several matters to the U.S. Department of Justice to obtain preliminary injunctions to immediately stop the sale of such products, including one product that when used as instructed was equivalent to industrial bleach.[12] The agency called its effort to root out fraudulent pandemic-related products "Operation Quack Hack."[13]

---

9. USDA's oversight of catfish products is a relatively recent development. *See* Michelle Johnson-Weider, *Muddying the Waters: Catfish Inspection Authority Transitions to the Food Safety and Inspection Service*, 13 J. Food L. & Pol'y 298 (2018).

10. *COVID-19: Update on Progress toward Safely Getting Back to Work and Back to School: Hearing before the Senate Committee on Health, Education, Labor, and Pension*, 116th Cong. (2020) (testimony of Stephen M. Hahn, Commissioner of Food and Drugs, FDA), https://www.help.senate.gov/imo/media/doc/HHS%20 COVID%20testimony%20June%2030%20HELP%20Committee%20-%20clear.pdf (emphasis added).

11. *Id.*

12. *Id.*; *COVID-19: An Update on the Federal Response: Hearing before the Senate Committee on Health, Education, Labor, and Pension*, 116th Cong. (2020) (testimony of Stephen M. Hahn, Commissioner, FDA) [hereinafter Second Testimony of Stephen M. Hahn], https://www.help.senate.gov/imo/media/doc/HHS%20testi mony%20COVID%20Senate%20HELP%209%2023%2020204.pdf.

13. Second Testimony of Stephen M. Hahn, *supra* note 12.

Throughout the COVID-19 pandemic, FDA consistently maintained that there is no evidence of food or food packaging being associated with transmission of SARS-CoV-2, COVID-19.[14] The U.S. Centers for Disease Control and Prevention (CDC) noted that the novel coronavirus is highly contagious, spreading primarily via respiratory droplets when individuals are in relatively close proximity to each other.[15] CDC directed "the more closely a person interacts with others and the longer that interaction, the higher the risk of COVID-19 spread."[16] While CDC identified that the novel coronavirus might also be spread if a person touches a surface or object with the virus on it then mechanically spreads the virus to his or her own eyes, mouth, or nose, the agency counseled that this was not the primary way the virus would spread.[17] FDA and USDA also issued a joint statement that "[c]onsumers should be reassured that we continue to believe, based on our understanding of currently available reliable scientific information, and supported by overwhelming international scientific consensus, that the foods they eat and food packaging they touch are highly unlikely to spread" the novel coronavirus.[18] Put simply, the agencies' announcements declared "the chances of infection by touching the surface of food packaging or eating food is considered to be extremely low."[19]

Given the realities of food production, the pandemic placed the food industry under significant strain, particularly in the agriculture and meat sectors. FDA noted that some components of the food production and manufacturing sector were "under stress" and that the agency was "identifying mitigation strategies."[20] Meanwhile, consumer buying patterns saw a significant shift from foodservice channels to retail grocery

---

14. *Id.*; *see also* FDA, *FDA Insight: Food Safety and COVID-19* ("I can assure the American people that the American food supply is safe and secure. There is no evidence of food or food packaging being associated with the transmission of COVID-19."), https://www.fda.gov/news-events/fda-insight/fda-insight-food-safety-and-covid-19 (last updated July 7, 2020).

15. CDC, *How COVID-19 Spreads*, https://www.cdc.gov/coronavirus/2019-ncov/prevent-getting-sick/how-covid-spreads.html (last updated Oct. 28, 2020).

16. *Id.*

17. *Id.*

18. Press Announcement, FDA, COVID-19 Update: USDA, FDA Underscore Current Epidemiologic and Scientific Information Indicating No Transmission of COVID-19 through Food or Food Packaging (Feb. 18, 2021), https://www.fda.gov/news-events/press-announcements/covid-19-update-usda-fda-underscore-current-epidemiologic-and-scientific-information-indicating-no.

19. *Id.*

20. Second Testimony of Stephen M. Hahn, *supra* note 12.

stores, challenging food supply chains to a breaking point. This severely decreased demand for foodservice and a rush to retail grocers initially led many consumers to see grocery store shelves empty.[21] During the pandemic, hospitality-oriented businesses, school and hospital cafeterias, and restaurants reported significant declines in traffic, often faced with capacity limitations, as individual consumers had stay-at-home mandates. Unfortunately, numerous food products destined for foodservice have packaging or other particular characteristics expressly designed for that channel, which limited their ability to be diverted quickly to retail grocers.[22] To address the situation, FDA and other agencies provided temporary flexibility in packaging and labeling requirements to help manufacturers and others divert products manufactured for foodservice and institutional use to retail grocery stores.[23] Federal agencies also identified that keeping workers safe while ensuring the continuity of operations for an essential food work force was a key priority area for pandemic response.

Mitigating Disruptions to the Food Supply Chain

The pandemic led to a seismic shift in how individuals obtained food. Online food shopping and delivery increased 500 percent.[24] Previously defined lines between foodservice and retail sales channels blurred, with regulators and market participants often scrambling to adapt, such as when market prices prompt farmers to plow fields rather than harvest their produce.[25] For online delivery, FDA noted that the agency needs to

---

21. FDA, *Food Safety and the Coronavirus Disease 2019 (COVID-19)* [hereinafter *Food Safety and COVID-19*] ("Why are there empty shelves at the local grocery store, while we see reports of food being dumped or crops being plowed under? (Posted Apr. 14, 2020)"), *archived at* https://perma.cc/84A3-WA3P (current as of July 29, 2020). *See also* FDA, *supra* note 14 ("[S]ome grocery store shelves may be temporarily short of certain items, mostly because customers are buying more than usual and not because there's less food.").

22. *See generally* Video: Why Millions of Potatoes Are Being Thrown Away during the Pandemic (Business Insider 2020), https://www.youtube.com/watch?v=ALtfQVbHtM0 (describing the plight of many potato crops designed specifically for foodservice).

23. Second Testimony of Stephen M. Hahn, *supra* note 12.

24. FDA, *supra* note 14.

25. *See* Adam Behsudi & Ryan McCrimmon, *Food Goes to Waste amid Coronavirus Crisis*, POLITICO, Apr. 5, 2020, https://www.politico.com/news/2020/04/05/food-waste-coronavirus-pandemic-164557; Michael Corkery & David Yaffe-Bellany, *"We Had to Do Something": Trying to Prevent Massive Food Waste*, N.Y. TIMES,

ensure such foods are produced, packaged, and transported safely, and that the "need for best practices in this area has been accelerated by the COVID-19 pandemic."[26]

In order to mitigate disruption to the food supply chain, FDA and USDA issued temporary policies providing actors in the supply chain with further flexibility. Among other things, FDA took the following steps. Each of the changes outlined in the table below were in effect only for the duration of the public health emergency. The agency has emphasized that they were temporary changes.

Temporary Policy	Detail
Allowing foods not labeled for retail sale to nonetheless be sold by restaurants and manufacturers in retail channels[27]	Many foods destined for foodservice are not sold with "nutrition facts" labeling. FDA exercised regulatory discretion to temporarily allow products to be sold at retail to consumers without this labeling so long as they have a statement of identity, an ingredient statement, the name and place of the business of the food manufacturer, packer, or distributor, a net quantity of contents, and required allergen information. FDA policy not only permits manufacturers to divert already packaged food, but also to produce new packages that may be sold for retail distribution.[28] USDA took similar measures for foods under its regulatory ambit.[29] For example, foods labeled for limited use only (i.e. "for school foodservice only"), which were already in commerce, could be distributed to retail outlets for repackaging and retail sale. On a temporary basis, USDA-regulated foods labeled for use in hotels, restaurants, and other institutions could be repackaged and would not require nutrition labeling.

May 2, 2020, https://www.nytimes.com/2020/05/02/business/coronavirus-food-waste-destroyed.html ("[W]e are not coming up with the supply-chain logistical solutions as quickly as produce is growing.").

26. FDA, *supra* note 14.

27. FDA, Temporary Policy regarding Nutrition Labeling of Certain Packaged Food during the COVID-19 Public Health Emergency: Guidance for Industry (2020), https://www.fda.gov/media/136469/download.

28. USDA has taken a similar position and also permits repackaging and relabeling of meat and poultry into consumer packages by retail outlets, without the mark of inspection. *See* USDA, FSIS Constituent Update: Temporary Allowances for Labels Going to Retail (2020), https://www.covid19businessguidance.com/wp-content/uploads/sites/40/2020/03/Foodservice-to-Retail-Update.pdf.

29. *See* Lawrence Reichman, *FDA and USDA Allow Food Packaged for Restaurants and Food Service to Be Repurposed for Consumers*, Perkins Coie, Mar. 27, 2020, https://www.covid19businessguidance.com/2020/03/fda-and-usda-allow

Temporary Policy	Detail
Permitting the sale of unlabeled shell eggs under certain conditions[30]	FDA does not intend to object to the sale of unlabeled eggs as long as the retail establishment provides certain information, such as a statement of identity and safe handling instructions.
Allowing product formulations to be adapted to account for supply shocks, without identifying all ingredients on the label[31]	Minor changes to product formulas, generally 2 percent or less, are temporarily permissible without updating the ingredient list. Manufacturers must ensure that the changes are indeed "minor" under the policy, and, for example, cannot change a characterizing ingredient or create an allergen issue.
Temporarily relaxing menu labeling requirements for chain restaurants[32]	Some dine-in operations are temporarily switching to a takeout-only format in response to the COVID-19 restrictions, which may involve creating new or modified online ordering portals and printed takeout or to-go menus for their stores. FDA does not intend to enforce menu labeling rules during the public health emergency related to the novel coronavirus.

In addition to availing themselves of these temporary regulatory flexibilities, manufacturers focused on keeping food safe during the pandemic. Manufacturers may also wish to consider updating their food

---

-food-packaged-for-restaurants-and-food-service-to-be-repurposed-for-consumers/; USDA, *supra* note 28.

30. Constituent Update, FDA, FDA Provides Temporary Flexibility regarding Packaging and Labeling of Shell Eggs Sold to Consumers by Retail Food Establishments during COVID-19 Pandemic (Apr. 3, 2020), https://www.fda.gov/food/cfsan-constituent-updates/fda-provides-temporary-flexibility-regarding-packaging-and-labeling-shell-eggs-sold-consumers-retail.

31. Constituent Update, FDA, FDA Announces Temporary Flexibility Policy regarding Certain Labeling Requirements for Foods for Humans during COVID-19 Pandemic (May 22, 2020), https://www.fda.gov/food/cfsan-constituent-updates/fda-announces-temporary-flexibility-policy-regarding-certain-labeling-requirements-foods-humans.

32. FDA, Temporary Policy regarding Nutrition Labeling of Standard Menu Items in Chain Restaurants and Similar Retail Food Establishments during the COVID-19 Public Health Emergency: Guidance for Industry (2020), https://www.fda.gov/media/136597/download.

safety plans to account for and mitigate the risks of the novel coronavirus.[33] This included assessing preventive controls, such as "kill steps," distribution holds, or sanitization procedures, that would mitigate risks of disease spread.[34] For manufacturers facing these kinds of choices, it was and remains important to document the efforts taken to evaluate and develop food safety plans in light of potential threats.[35]

Food Work Environments

During the pandemic, food and agriculture workers were designated as "essential workers," protecting the critical infrastructure required to maintain a robust national food supply.[36] But a sick work force during a short growing season could be devastating, especially to smaller producers in addition to the individual workers and their families. Essential workers in the nation's grocery stores and on its farms are risking their lives to make sure that food is available for the American people.

As with any employer, food employers must assess their duties under laws governing employment and workplace safety, including health and safety considerations under the Occupational Safety and Health Act, discrimination and leave considerations, and employee privacy considerations.[37] Especially important for food businesses are healthy hygiene practices, including frequent handwashing, avoiding contact with one's face, and maintaining social distancing. Food employers should also consider written protocols and policies to address, among other things, when illness should be disclosed, what leave benefits are available, when tests might be provided, and how the employer will maintain privacy.

In August 2020, the Occupational Safety and Health Administration (OSHA) and FDA issued a joint checklist entitled *Employee Health and Food Safety Checklist for Human and Animal Food Operations during*

---

33. Lawrence Reichman, *FDA Issues COVID-19 Food Safety Guidance for Food Manufacturers*, Perkins Coie, Mar. 23, 2020, https://www.perkinscoie.com/en/news-insights/fda-issues-covid-19-food-safety-guidance-for-food-manufacturers.html, *archived at* https://perma.cc/N8SH-UXS7.

34. *Id.*

35. *Id.*

36. *See* California State Coronavirus Response, Essential Workforce (2020), https://files.covid19.ca.gov/pdf/EssentialCriticalInfrastructureWorkers.pdf.

37. Sarah E. Flotte et al., *Coronavirus (COVID-19) and the Workplace: Practical Considerations and Tips for US Employers*, Perkins Coie, Feb. 28, 2020, https://www.perkinscoie.com/en/news-insights/coronavirus-covid-19-and-the-workplace-practical-considerations-and-tips-for-us-employers.html.

the COVID-19 Pandemic.[38] Among other things, the agencies suggested developing a COVID-19 assessment and control plan, a workplace coordinator, and a clear point of contact to communicate reports of recent symptoms and to consult with employees about presumptive cases of COVID-19.[39]

FDA did not anticipate that food products would need to be recalled or withdrawn from the market because of COVID-19, "as there is currently no evidence to support the transmission of COVID-19 associated with food or food packaging."[40] Food employers are often required to maintain clean and sanitized facilities and food contact surfaces, which helps control the risk that ill workers might spread a disease, regardless of the type of virus or bacteria involved.[41]

FDA issued a series of questions and answers regarding securing food workplaces during the pandemic. Among other things, these include the select questions and quoted answers included in the table below.

Question	Answer
What steps do I need to take to clean the facility/equipment to prevent the spread of COVID-19?[42]	FDA-regulated food manufacturers are required to follow Current Good Manufacturing Practices (CGMPs) and many have food safety plans that include a hazards analysis and risk-based preventive controls. CGMPs and food safety plans have requirements for maintaining clean and sanitized facilities and food contact surfaces.[43] [FDA encourages] coordination with local health officials for all businesses so that timely and accurate information can guide appropriate responses in each location where their operations reside. Food facilities may want to consider a more frequent cleaning schedule.[44]

---

38. FDA & OSHA, EMPLOYEE HEALTH AND FOOD SAFETY CHECKLIST FOR HUMAN AND ANIMAL FOOD OPERATIONS DURING THE COVID-19 PANDEMIC (2020), https://www.fda.gov/media/141141/download, *archived at* https://perma.cc/576H-93AK.
39. *Id.* at 2.
40. *Food Safety and COVID-19, supra* note 21.
41. *Id.*
42. *Id.* (follow "What steps do I need to take to clean the facility/equipment to prevent the spread of COVID-19? (Posted March 17, 2020)").
43. *Id.*
44. *Id.*

Question	Answer
How do I maintain social distancing in my food production/processing facility and food retail establishment where employees typically work within close distances?[45]	To prevent spread of COVID-19, CDC is recommending individuals employ social distancing or maintaining approximately 6 feet from others, when possible. In food production/processing facilities and retail food establishments, an evaluation should be made to identify and implement operational changes that increase employee separation. However, social distancing to the full 6 feet will not be possible in some food facilities.
	The risk of an employee transmitting COVID-19 to another is dependent on distance between employees, the duration of the exposure, and the effectiveness of employee hygiene practices and sanitation. When it's impractical for employees in these settings to maintain social distancing, effective hygiene practices should be maintained to reduce the chance of spreading the virus. . . .
	IMPORTANT: Maintaining social distancing in the absence of effective hygiene practices may not prevent the spread of this virus. Food facilities should be vigilant in their hygiene practices, including frequent and proper hand-washing and routine cleaning of all surfaces.
	Because the intensity of the COVID-19 outbreak may differ according to geographic location, coordination with state and local officials is strongly encouraged for all businesses so that timely and accurate information can guide appropriate responses in each location where their operations reside.
	Sick employees should follow the CDC's [guidance on] What to do if you are sick with coronavirus disease 2019 (COVID-19).[46]

---

45. *Id.* (follow "How do I maintain social distancing in my food production/processing facility and food retail establishment where employees typically work within close distances? (Updated April 5, 2020)").

46. *Id.*

Question	Answer
A worker in my food production/processing facility/ farm has tested positive for COVID-19. What do I need to do to protect other employees and continue operations?[47]	To ensure continuity of operations, CDC advises that critical infrastructure workers may be permitted to continue work following potential exposure to COVID-19, informed by the risk assessment of the workplace that accounts for COVID-19 mitigations already in place, provided they remain symptom-free and additional precautions are taken to protect them and the community.
	Sick workers should stay home or go home if they develop symptoms during the work day. For workers potentially exposed to someone with COVID-19, employers should • Check temperatures and assess symptoms of workers, ideally before entering the facility or operation. • If no fever (>100.4 F) or COVID-19 symptoms are present, workers should self-monitor for onset of symptoms during their shift.
	As a general good practice for all workers, employers should • Encourage workers to use an employer-approved face mask or cloth face covering at all times while in the workplace. • Ensure that workers can practice social distancing or employ engineering solutions if that is not possible. • Make available facilities and materials for worker hygiene so workers can practice CDC recommended handwashing. • Clean and disinfect workplaces/stations at frequent intervals.[48]

---

47. *Id.* (follow "A worker in my food production/processing facility/farm has tested positive for COVID-19. What do I need to do to protect other employees and continue operations? (Posted July 17, 2020)").

48. *Id.*

Question	Answer
If a worker in my food processing facility has tested positive for COVID-19, should I test the environment for the SARS-CoV-2 virus?[49]	Currently there is no evidence of food or food packaging being associated with transmission of COVID-19. Facilities are required to use personnel practices that protect against contamination of food, food contact surfaces and packaging and to maintain clean and sanitized facilities and food contact surfaces. Although it is possible that the infected worker may have touched surfaces in your facility, FDA-regulated food manufacturers are required to follow Current Good Manufacturing Practices (CGMPs). Maintaining CGMPs in the facility should minimize the potential for surface contamination and eliminate contamination when it occurs. With the detection of the coronavirus in asymptomatic people and studies showing survival of coronavirus on surfaces for short periods of time, as an extra precaution, food facilities may want to consider a more frequent cleaning and sanitation schedule for high human contact surfaces.[50]
If a worker in my food processing facility/farm has tested positive for COVID-19, should I close the facility? If so, for how long?[51]	Food facilities need to follow protocols set by local and state health departments, which may vary depending on the amount of community spread of COVID-19 in a given area. *These decisions will be based on public health risk of person-to-person transmission—not based on food safety.*[52]

---

49. *Id.* (follow "If a worker in my food processing facility has tested positive for COVID-19, should I test the environment for the SARS-CoV-2 virus? (Updated July 17, 2020)").

50. *Id.*

51. *Id.* (follow "If a worker in my food processing facility/farm has tested positive for COVID-19, Should I close the facility? If so, for how long? (Updated July 17, 2020)").

52. *Id.* (emphasis added).

Question	Answer
Do I need to recall food products produced in the facility during the time that the worker was potentially shedding virus while working?[53]	[FDA does] not anticipate that food products would need to be recalled or be withdrawn from the market because of COVID-19, as there is currently no evidence to support the transmission of COVID-19 associated with food or food packaging. Additionally, facilities are required to control any risks that might be associated with workers who are ill regardless of the type of virus or bacteria. For example, facilities are required to maintain clean and sanitized facilities and food contact surfaces.[54]

Especially in a crisis, food manufacturing, processing, and retail are critical pieces of infrastructure. Securing food work environments helps not only protect this critical infrastructure, but also keeps workers and their families safe.

Addressing Food Insecurity

As families continued to wrestle with the economic consequences of COVID-19 stemming from individual health issues and job losses, food banks saw an unprecedented level of need.[55] According to anti-hunger advocate Joel Berg, "the pandemic only worsened a massive, pre-existing hunger problem in the U.S."[56] One in six households with children reported food insecurity, and racial and ethnic disparities have deepened, with Black and Latinx households seeing considerably higher rates of food insecurity than other households.[57] Nearly 14 million

---

53. *Id.* (follow "If a worker in my food processing facility/farm has tested positive for COVID-19, Should I close the facility? If so, for how long? (Updated July 17, 2020)").

54. *Id.*

55. Helena Bottemiller Evich, *"There's Only So Much We Can Do": Food Banks Plead for Help*, POLITICO, June 8, 2020, https://www.politico.com/news/2020/06/08/food-banks-plead-for-help-306492.

56. Charles Platkin, *Interview with Joel Berg, NYC's Hunger Free Advocate*, N.Y. CITY FOOD POL'Y CENTER, Mar. 9, 2021, https://www.nycfoodpolicy.org/interview-with-joel-berg-nycs-hunger-free-advocate/.

57. Helena Bottemiller Evich, *"Unprecedented": 1 in 6 Households with Children Report Kids Are Not Getting Enough to Eat*, POLITICO, July 9, 2020, https://www.politico.com/news/2020/07/09/food-insecurity-children-355425; Evich, *supra* note 7.

children lived in households experiencing food insecurity during the pandemic, which is 2.7 times the amount that did at the peak of the Great Recession in 2008.[58]

USDA coordinates a number of federal nutrition programs, including the Supplemental Nutrition Assistance Program (SNAP) (formerly called the Food Stamp Program), the Special Supplemental Nutrition Program for Women, Infants, and Children (WIC), and the nation's school meal programs. During the pandemic, there have been several significant changes with regard to food assistance.[59] For example, Congress created a pandemic food benefit program that provides money for children who would otherwise receive free or reduced-price meals but cannot because of school closures.[60] In addition, emergency allotments for food stamps have allowed those receiving SNAP benefits at any benefit level to receive the maximum benefit allowable for their household size.[61] Generally, SNAP benefits vary across income levels and household size, and the new emergency allotments increased benefit levels for many SNAP recipient households. In addition, online purchasing using SNAP benefits has accelerated, with a pilot program now running in about 40 states.[62] This pilot may lead to increased use of online SNAP purchasing after the crisis as well.

The Biden Administration announced new measures to broaden food assistance benefits.[63] The new administration declared that pandemic food benefits to children would increase 15 percent to "tackle the serious problem of child food insecurity."[64] With the passage of the American Rescue Plan, USDA has announced it would increase SNAP benefits by 15 percent, or $28 per person between March and September 2021.[65] The agency estimates this increase will provide more than 100 additional

---

58. Bauer, *supra* note 5.
59. *See* Tobin, *supra* note 4.
60. *See* Tommy Tobin, *Report: Pandemic Food Benefits Work to Address Food Insecurity*, FORBES, Aug. 10, 2020, https://www.forbes.com/sites/tommytobin/2020/08/10/hamilton_project_report_p-ebt_pandemic_food/.
61. *See* Tommy Tobin, *Expanded Emergency Food Stamp Benefits Blocked by Federal Court*, FORBES, June 19, 2020, https://www.forbes.com/sites/tommytobin/2020/06/19/covid-food-stamp-usda-allotments-blocked/?sh=5e8881492cac.
62. *Id.*
63. Press Release, USDA Food and Nutrition Service, Biden Administration Expands P-EBT to Benefit Millions of Low-Income and Food Insecure Children during Pandemic (Jan. 22, 2021), https://www.fns.usda.gov/news-item/usda-001521.
64. *Id.*
65. Press Release, USDA, USDA Increases SNAP Benefits Up to $100 per Household with Funding from American Rescue Plan (Mar. 22, 2021), https://

dollars per month for a family of four. The agency is also revising the Thrifty Food Plan to better reflect the costs of a healthy basic diet on a more permanent basis. This review might find that the current Thrifty Food Plan, upon which SNAP benefits are based, is inadequate to afford a healthy basic diet, and should therefore be revised upwards.

What Is Ahead

As of this writing, the story is still unwritten about the end of the pandemic and its economic realities. Companies, consumers, and regulators continue to adapt to new realities of the COVID-19 pandemic, and some of the changes implemented during this emergency period may lead to long-term shifts. For example, customer safety, both real and perceived, will be an important issue for restaurants gearing up to reopen when the pandemic eventually wanes. The increase in online shopping for groceries might well lead to regulations requiring online labeling and disclosures similar to those required on product labels, which is already required for so-called "distance selling" in the European Union.[66]

A leading trade association, the Consumer Brands Association, issued a report about changes to advance smarter consumer packaged goods product regulation during and after the pandemic.[67] The trade association found that the pandemic created unprecedented challenges for food and consumer packaged goods companies, and many "crisis-driven actions have since proven to be as safe and even more efficient at achieving the goals of pre-crisis public policy and should be made permanent."[68] The group targeted recommendations in transportation, inspections, manufacturing, and labeling. For example, the Consumer Brands Association advocated for the expansion of digital disclosure of product information and maintaining flexibility for food labeling to facilitate the transfer of products between the food retail and foodservice channels.

*****

---

www.usda.gov/media/press-releases/2021/03/22/usda-increases-snap-benefits-100-household-funding-american-rescue.

66. *See* Council Regulation 1169/2011 on the Provision of Food Information to Consumers, art. 14, 2011 O.J. (L 304) 18, https://eur-lex.europa.eu/legal-content/EN/TXT/HTML/?uri=CELEX:32011R1169&from=EN.

67. Consumer Brands Association, Applying Lessons Learned from the COVID-19 Crisis: Ten Critical Changes to Advance Smarter CPG Industry Regulatory Policy (2020), https://consumerbrandsassociation.org/wp-content/uploads/2020/06/CBA_Top10.pdf.

68. *Id.*

The economic realities of the COVID-19 pandemic will continue to ripple for years to come. Already, the novel coronavirus has had considerable human cost. The safety and security of the food supply have never been more essential, and the industry rose to meet the challenge. Regulators at FDA, USDA, and elsewhere provided regulatory flexibility to assist industry and individual consumers during this time of unprecedented challenge.

Chapter 10

Intellectual Property

Anne W. Glazer

Introduction

Intellectual property comes in four types: patent, trade secret, trademark, and copyright. A single work or a single product may receive more than one type of intellectual property protection, but the principles and rules for each type are distinct. A logo may be protected as a trademark of the brand owner by virtue of its use in selling products, and also protected as a copyright of the logo designer by virtue of its affixation in an electronic file. A product design or package, as trade dress, may be protected as a trademark of the brand owner by virtue of its public recognition, and also protected as a design patent by virtue of its registration with the U.S. Patent and Trademark Office (USPTO).

In the food world, utility patents often protect agricultural and industrial sources and processes, including seeds, plants, and manufacturing methods. The law of trade secrets protects the unpatented know-how and recipes that go into making food, as well as the sensitive pricing, supply, customer, and other confidential information that every business guards from competitors.

Trademarks protect brand identities at every step in the supply chain, but especially at the consumer level where the world's most valuable brands include Coca-Cola, McDonald's, Walmart, Budweiser, Nescafé,

Pepsi, Starbucks, and Frito-Lay.[1] In advertising, the law of trademarks enables a marketer to stop competitors from using confusingly similar brand names and logos, while the law of copyrights enables a photographer or designer to demand payment for the use of his or her work.

A practitioner of intellectual property law is called upon to assist the food industry at each stage in the life cycle of a business, brand, or product, often including:

- Selection, clearance, and registration of a company name, "house mark," line name, product name, advertising tag line, or logo
- Assessment of patentability and prosecution of patents
- Nondisclosure agreements (NDAs) protecting information shared in the evaluation of a potential business relationship or in the course of a business relationship
- Employee confidentiality and proprietary rights agreements protecting business confidential information and clarifying the scope of intellectual property ownership
- Work for hire agreements securing copyrights created by consultants and independent contractors in "creative" works such as photographs, label designs, advertisements, and websites
- Licensing software and other intellectual property to or from others
- Monitoring the marketplace for infringers and enforcing rights against them
- Defending against intellectual property "cease and desist" and licensing demands
- Handling intellectual property due diligence, representations and warranties, and ancillary agreements in mergers and acquisitions (M&As)

In addition, intellectual property practitioners are often called upon to assist with adjacent matters, including Internet domain names, rights of privacy and publicity, and various commercial contracts.

---

1. Marty Swant, *The World's Most Valuable Brands 2020*, FORBES, July 27, 2020, https://www.forbes.com/the-worlds-most-valuable-brands/.

Patent

Patent law protects inventions.[2] A patent is arguably the most powerful form of intellectual property protection because it grants the owner the right to exclude others from making or using the invention, regardless of industry, context, intent, or knowledge. Patents can be independently monetized, and infringement cases can bring big damages awards. Patents are also the costliest form of intellectual property to obtain. Regardless, it is generally wise to consider potential patenting upon first contact with an inventive innovation. Time is of the essence in deciding whether to pursue a patent.

Patenting plays an important role in agriculture and industrial processing. Foods and food production methods are generally eligible subject matters for patenting. However, patents are rarely available to ordinary food manufacturers for the protection of their recipes or methods, which generally rely on trade secret protection.

In order to be patentable, an invention must be more than an abstract idea.[3] For the protection of an idea, turn to the law of trade secrets. A patent must describe the invention specifically enough to enable a person skilled in the art to make and use it.[4]

The Patent Definition

Whoever invents or discovers any new and useful process, machine, manufacture, or composition of matter, or any new and useful improvement thereof, may obtain a utility patent therefor, subject to the conditions and requirements of the Patent Act.[5] A patent will be denied, however, if the invention would have been obvious to a person having ordinary skill in the art to which the invention pertains.[6] Thus, the Patent Act sets out the three requirements that must be met in order to obtain a utility patent: the invention must be new, useful, and nonobvious.

Plant patents exist separately in a corner of the patent practice. A plant patent may be granted to whoever invents or discovers and asexually reproduces a distinct and new plant variety, other than a tuber-propagated plant or a plant found in an uncultivated state.[7] Growers often

---

2. 1 DONALD S. CHISUM, CHISUM ON PATENTS 1 (2021) [hereinafter CHISUM ON PATENTS].
3. Alice Corp. Pty v. CLS Bank Int'l, 573 U.S. 208 (2014).
4. 35 U.S.C. § 112(a).
5. *Id.* § 101.
6. *Id.* § 103.
7. *Id.* § 161.

pay royalties to breeders for the right to reproduce patented trees and plants.[8] A plant may also be eligible for a traditional utility patent, as in the field of genetically modified organisms.[9]

While an invention can be protected as a trade secret unless and until it is patented, the issuance of a patent is incompatible with trade secret protection. A patent is a public record that represents a bargain: it discloses how to make and use the invention in return for the right to exclude others during the limited patent term.

On-Sale Bar

The "on-sale bar" encourages early discussions about patenting. As part of the novelty requirement, an invention cannot be patented if it has already been "patented, described in a printed publication, or in public use, on sale, or otherwise available to the public."[10] There is a one-year grace period for a disclosure made by the inventor or an associated party, including a sale or offer to sell, among other exceptions.[11] This is a complex body of law. Further, there is no grace period in key foreign countries, where the ability to patent can be lost upon disclosure of an invention.

Many inventors manage the process and budget by first filing a provisional application, which can be a simple and broad description of the invention and therefore costs relatively little. This filing stops the ticking of the "on-sale bar" clock. After filing a provisional application, the applicant has one year in which to file a full application. In the meantime, the applicant may continue to improve and test the invention.

Competent client representation in patent matters requires scientific and technical knowledge and skill as well as legal knowledge and skill.[12] The practice of patent law requires special examination and registration with the USPTO. According to USPTO rules, only registered patent practitioners may ethically advise clients in contemplation of drafting or filing

---

8. *See* Lela Nargi, *Intellectual Property and Trademark Protections for Fruit-Growing Plants Are on the Rise—And So Are the Lawsuits*, Counter, July 28, 2020, https://thecounter.org/intellectual-property-trademark-ip-law-fruit-cosmic-crisp-cotton-candy-grapes/.

9. *See* Bowman v. Monsanto Co., 569 U.S. 278 (2013) (holder of utility patent for genetically altered soybean seed successfully sued farmer for planting and harvesting seed).

10. 35 U.S.C. § 102(a).

11. *Id.* § 102(b)(1).

12. USPTO Rules of Professional Conduct § 11.101 (2013).

a patent application.[13] Therefore, in light of the unforgiving on-sale bar, it is best to quickly refer a hopeful inventor to a registered patent attorney.

Utility Patents

The requirements of novelty and nonobviousness have been tested in the food industry. For example, Procter & Gamble (P&G) obtained a patent claiming the recipe and process for making a cookie that is crispy on the outside and chewy on the inside. When P&G sued Keebler, Nabisco, and Frito-Lay for infringement, the defendants argued that the patent was anticipated by a recipe published in a 1968 cookbook, and therefore invalid. In a partial summary judgment, the court held that the 1968 publication defeated P&G's novelty claim.[14]

Design Patents

A utility patent protects the way an item is made or the way it works, while a design patent protects the way it looks. Whoever invents a new, original, and ornamental design for an article of manufacture may obtain a design patent therefor, subject to the conditions and requirements of the Patent Act.[15] Eligible designs may include surface ornamentation on a product, the configuration of all or part of a product, or a combination of both.[16]

Design patents may be granted for food packaging or, occasionally, a food. The shape of the original Coca-Cola bottle was the subject of a design patent. Other examples of issued design patents include Cold Stone Creamery's ice cream cake (D571526S1), Good Humor Breyer's Viennetta ice cream cake (D486951S1), and Condor Snack Company's pretzel shaped like a peace sign (D423184S).

In a design patent, the drawing is the one and only claim. A design patent grants the owner the right, for a term of 14 years, to exclude others from making articles that, to the ordinary observer, look substantially the same as the patented design when considered in context of prior designs, inducing the observer to purchase one supposing it to be the other.[17] The scope of protection is thus rather narrow and can be defeated by relatively small changes in the design. On the other hand, a design patent does not

---

13. 37 C.F.R. § 11.5(b)(1).
14. Procter & Gamble Co. v. Nabisco Brands, Inc., 711 F. Supp. 759, 760 (D. Del. 1989).
15. 35 U.S.C. § 171.
16. *In re* Zahn, 617 F.2d 261 (C.C.P.A. 1980).
17. Lanard Toys Ltd. v. Dolgencorp LLC, 958 F.3d 1337 (Fed. Cir. 2020).

cost much to obtain and can be a useful enforcement tool, particularly against copycats.

The Effect of a Patent

Unlike the other principal forms of intellectual property, a patent does not exist until it is issued by the government. A "patent pending" notice merely means a patent application has been filed. It does not indicate the existence of any intellectual property protection.

A patent owner has the right to exclude others from making or using the patented invention during the limited term of the patent, which is 20 years for a utility patent.[18] In contrast to copyright, a patent owner need not show that the invention was copied in order to state a claim for infringement.

The law of patent infringement is complex and patent disputes can be extremely costly. As a starting point, the court must construe the patent's claims, which are the operative language that defines the metes and bounds of the patent. A patent is literally infringed if the claim "reads on" the allegedly infringing process or product. If the patent is not literally infringed, infringement may still be found under the doctrine of equivalents, if the accused process or product has elements that are equivalent to each claimed element of the invention.[19]

Patent Infringement Remedies

A court may award money damages and an injunction in case of patent infringement.[20] Damages are usually measured in terms of lost profits or a reasonable royalty.[21] In cases of willful infringement, a court may award treble damages and attorneys' fees.[22]

Trade Secret

In practice, few food products or processes are eligible for patent protection. As we will see, copyright cannot be used to protect food or other useful items, except for narrow situations such as two-dimensional designs on cakes. Trademark law protects brand signifiers, but it rarely protects a feature of the product itself. What is left? Trade secret protection.

---

18. 5 CHISUM ON PATENTS, *supra* note 2, § 16.02.
19. Warner-Jenkinson Co. v. Hilton Davis Chem. Co., 520 U.S. 17 (1997).
20. 35 U.S.C. § 283.
21. 7 CHISUM ON PATENTS, *supra* note 2, § 20.01.
22. 35 U.S.C. §§ 284, 285; Halo Elecs., Inc. v. Pulse Elecs., Inc., 136 S. Ct. 1923 (2016).

Trade secrets play a major role in food manufacturing and food services, particularly in relation to formulas and recipes, equipment, processes, sources, relationships, and pricing. The formula for Coca-Cola is arguably the most valuable trade secret of all time. Not only are extensive security measures taken at Coca-Cola's headquarters, reportedly the only two people who know the whole formula fly on separate planes in case of an accident.[23]

Keeping Trade Secrets Secret

Unlike the other forms of intellectual property, trade secrets cannot be "certificated" and they have, until recently, been governed mostly by state laws with reference to the Restatement of Torts. The Restatement continues to be useful to courts, but the law of trade secrets is now governed overwhelmingly by statute.[24] Every state except New York has now adopted either a version of the Uniform Trade Secrets Act (UTSA) or something similar to it. New York has no general trade secrets statute; instead, it relies on the common law, including the Restatement (First) of Torts.[25]

Further, since the passage of the Defend Trade Secrets Act of 2016 (DTSA), federal law now plays a greater role in trade secrets jurisprudence. The DTSA is codified as part of the Economic Espionage Act, which has long provided criminal and civil penalties for trade secret misappropriation in limited circumstances.[26] The DTSA provides a federal private right of action and a suite of civil remedies for trade secret misappropriation.[27] As a result, it is now common for trade secret owners to assert misappropriation claims under both the UTSA and the DTSA.

The Trade Secret Definition

Under the UTSA, a "trade secret" is any information, including a formula, pattern, compilation, program, device, method, technique, or process, that (1) derives actual or potential independent economic value from not being generally known to, and not being readily ascertainable by proper means by, other persons who can obtain economic value from

---

23. Zachary Crockett, *The Botched Coca-Cola Heist of 2006*, HUSTLE, Apr. 28, 2018, https://thehustle.co/coca-cola-stolen-recipe.
24. 1 ROGER M. MILGRIM & ERIC E. BENSEN, MILGRIM ON TRADE SECRETS § 1.01 (2021) [hereinafter MILGRIM ON TRADE SECRETS].
25. *Id.*
26. 18 U.S.C. § 1831 *et seq.*
27. *Id.* § 1836.

its disclosure or use, and (2) is the subject of efforts that are reasonable under the circumstances to maintain its secrecy.[28] The DTSA definition is very similar.[29]

Under both the UTSA and the DTSA, there are three elements in the trade secret definition. First, the subject information must actually be *secret*. Specifically, it must not be generally known to the public or to other persons who can obtain economic value from its disclosure or use.[30] Most states and the DTSA also provide that the information must not be readily ascertainable through proper means by those who can gain economic value from its disclosure or use.[31]

Second, the information must derive actual or potential economic value from being secret.[32]

Third, the information must be the subject of efforts that are reasonable under the circumstances to keep it a secret.[33] In other words, it is not enough to show that the information is secret. A claimant must also demonstrate that it has made reasonable efforts to maintain secrecy.

A trade secret may be any type of information that meets the definitional test. Importantly for the protection of recipes and manufacturing processes, a trade secret may be a compilation of known elements.[34] In *Bimbo Bakeries*, the plaintiff alleged that U.S. Bakery misappropriated its trade secrets by hiring its employee and modifying its "granny-style" bread based on confidential methods learned from that employee. U.S. Bakery argued that the process was a mere compilation of four elements—ingredients and production steps—that were common knowledge in the trade. A jury found that the *combination* of elements was a protected trade secret. The jury awarded damages on the claim in the amount of $2,105,256, and the court added exemplary damages against U.S. Bakery in the amount of $789,471.[35]

---

28. UNIF. TRADE SECRETS ACT § 1(4) (1986); *see also* CAL. CIV. CODE § 3426.1(d).
29. *See* 18 U.S.C. § 1839.
30. *Id.* § 1839(3); CAL. CIV. CODE § 3426.1(d)(1).
31. *E.g.*, 18 U.S.C. § 1839(3); FLA. STAT. § 688.002(4); 12 PA. CONS. STAT. § 5302; TEX. CIV. PRAC. & REM. CODE § 134A.002(6).
32. *E.g.*, 18 U.S.C. § 1839(3); CAL. CIV. CODE § 3426.1(d)(1).
33. *E.g.*, 18 U.S.C. § 1839(3); CAL. CIV. CODE § 3426.1(d)(2).
34. *E.g.*, Bimbo Bakeries USA, Inc. v. Sycamore, 372 F. Supp. 3d 1291 (D. Utah 2019).
35. *See* Bimbo Bakeries USA, Inc. v. Sycamore, No. 2:13-cv-00749-DN-DBP, 2018 U.S. Dist. LEXIS 54556 (D. Utah Mar. 29, 2018).

Misappropriation

Misappropriation involves the acquisition of a trade secret by improper means or the knowing unauthorized disclosure or use of a trade secret that was acquired by accident or mistake. The UTSA and the DTSA define misappropriation similarly:

> (1) acquisition of a trade secret by a person who knows or has reason to know it was acquired by improper means; or
> (2) disclosure or use of a trade secret by a person who either:
> (A) used improper means to acquire it;
> (B) knew or had reason to know that it was:
> (i) acquired from a person who used improper means to acquire it
> (ii) acquired under circumstances giving rise to a duty to maintain secrecy or limit use, or
> (iii) acquired from another person who had a duty to maintain secrecy or limit use; or
> (C) before a material change of position, knew or had reason to know that it was a trade secret acquired by accident or mistake.[36]

"Improper means" includes theft, bribery, misrepresentation, breach of duty to maintain secrecy (or inducement thereof), or espionage.[37]

Remedies

A court may enjoin actual or threatened trade secret misappropriation.[38] The DTSA also provides for civil seizure in extraordinary circumstances.[39]

Damages for trade secret misappropriation may include both actual losses and any unjust enrichment that is not accounted for in computing actual losses. The damages may be measured by any appropriate method, including a reasonable royalty for the unauthorized disclosure or use.[40] The statutes also provide for exemplary damages of up to two times the award, plus attorneys' fees, in cases of willful and malicious misappropriation.[41]

---

36. Cal. Civ. Code § 3426.1; 18 U.S.C. § 1839(5)(B).
37. Cal. Civ. Code § 3426.1(a); 18 U.S.C. § 1839(6).
38. Cal. Civ. Code § 3426.2; 18 U.S.C. § 1836(b)(3)(A).
39. 18 U.S.C. § 1836(b)(2).
40. Cal. Civ. Code § 3426.3; 18 U.S.C. § 1836(b)(3)(B).
41. Cal. Civ. Code §§ 3426.3, 3426.4; 18 U.S.C. § 1836(b)(3)(C)–(D).

Federal criminal prosecution is also a possibility. In 2007, a former Coca-Cola administrator was sentenced to eight years in prison—more than prosecutors requested—for stealing documents and a vial of product under development and offering them to Pepsi.[42] Her accomplice received a five-year sentence.[43]

NDAs

Most business lawyers touch trade secret issues from time to time, thanks to the ubiquity of NDAs. Not only are such agreements important for maintaining secrecy, they are also key to demonstrating that a party has made reasonable efforts to maintain secrecy.

NDAs are so common and relatively simple that it can be easy to overlook serious issues with their wording. In developing an NDA, the following should be considered:

- The definition of "protected information" is often a broad category (e.g., "all non-public or confidential information") followed by a list of examples.
- A marking requirement may or may not be advisable. Will the client have the discipline and means to mark all its confidential information? How will the parties handle information that is disclosed orally or by observation, such as on a plant floor?
- Under what circumstances may the receiving party disclose the information to a third party, and to whom? Typically, the agreement will at least require that any third-party recipient must be bound by an obligation to keep the information confidential.
- In addition to prohibiting the disclosure of confidential information, an NDA should generally prohibit its *use* for anything other than the articulated purpose.
- The agreement may run indefinitely or have a definite term.
- Separately, the agreement should clarify how long a party's obligation to protect the information will survive. For example, it may survive for a term of years or for as long as the protected information remains confidential or a trade secret.

---

42. *Prison Looms for Coca-Cola Recipe Thief*, BEVERAGEDAILY.COM, May 23, 2007, https://www.beveragedaily.com/Article/2007/05/24/Prison-looms-for-Coca-Cola-recipe-thief.

43. *Id.*

Employees Coming and Going

The lion's share of trade secret litigation involves departed employees.

State laws generally impose duties of loyalty and confidentiality that prohibit an employee from using or disclosing the employer's confidential information to the employer's detriment.[44] However, these rules vary widely in their application.

Therefore, it is generally wise for a company to have each of its employees sign a confidentiality and proprietary rights agreement that (1) calls their attention to the importance of protecting confidential information, (2) expresses the obligation to protect all company confidential information and not only statutory trade secrets, and (3) provides a reasonable post-termination covenant to continue maintaining confidentiality. Notably, evidence of such agreements is routinely requested and expected as part of buyer due diligence in M&A transactions.

In addition, a company's employee handbook should emphasize the importance of trade secret protection and point out the duty of every employee to maintain confidentiality. Every employee should be required to sign a confirmation that they have received the handbook.[45]

The theory of "inevitable disclosure" is the subject of much discussion. In the leading case, *PepsiCo, Inc. v. Redmond*,[46] Mr. Redmond was a top marketing executive at PepsiCo and bound by a confidentiality agreement when he left to work for Quaker Oats Company, specifically the Gatorade and Snapple brands. PepsiCo demonstrated that Redmond had extensive and intimate knowledge about its pricing, distribution, and marketing strategies in the sports and "new age" drink categories, which he could not help but draw upon in his work for Quaker Oats. Notably, the court found that Mr. Redmond and his contact at Quaker Oats had demonstrated a lack of candor, and that Mr. Redmond had taken certain documents with him when he left PepsiCo.[47] The Seventh Circuit affirmed an

---

44. 1 MILGRIM ON TRADE SECRETS, *supra* note 24, § 5.02; Annotation, *Implied Obligation of Employee Not to Use Trade Secrets or Confidential Information for His Own Benefit or That of Third Persons after Leaving the Employment*, 165 A.L.R. 1453, 1454 (1946).

45. *See, e.g.*, Barilla Am., Inc. v. Wright, No. 4-02-CV-90267, 2002 WL 31165069 (S.D. Iowa July 5, 2002) (quoting handbook as the key evidence that a departing employee was aware of his duty not to disclose confidential information and of the sensitivity of the information in the plaintiff's pasta manufacturing facility).

46. 54 F.3d 1262 (7th Cir. 1995).

47. *Id.*

order enjoining Redmond from assuming his responsibilities at Quaker Oats for some period of time.[48]

California has rejected the so-called inevitable disclosure doctrine on the ground that it has the same effect as a post-employment covenant not to compete, which California law forbids.[49] Other courts have declined to give it independent force. For example, in *Del Monte Fresh Produce Co. v. Dole Food Co.*,[50] the court refused to apply the doctrine in the absence of evidence of actual or threatened misappropriation. When courts do rely on an inevitable disclosure theory, it is generally in cases where the new employer is in direct competition with the complainant and the employee is shown to possess significant trade secret information.

Trademark

Trademark law protects the symbols and signs that identify a brand. Trademarks differ from other forms of intellectual property in that they are inseparable from business goodwill. No one can "own" a word, not even APPLE, though some famous coined marks like GOOGLE and COCA-COLA come close. A trademark does not consist of a word, symbol, or other device in itself; rather, it is defined in connection with the particular goods or services on which it is used.

Also, unlike patent or copyright, trademark is not mentioned in the U.S. Constitution and is not solely a creature of federal law. The framers saw fit to provide for the interests of patent inventors and copyright authors "[t]o promote the [p]rogress of [s]cience and useful [a]rts."[51] Trademark, in contrast, is part of the law of unfair competition.[52] As such, it is interested not only in protecting trademark owners and marketers, but also in protecting consumers and purchasers from confusion.[53]

In the United States, a trademark comes into existence when it is used to sell products or services. Thus, U.S. trademark rights are fundamentally based on the use of a mark, not on registration. An unregistered trademark may be fully enforceable, even in federal court.[54] Furthermore,

---

48. *Id.*
49. Whyte v. Schlage Lock Co., 125 Cal. Rptr. 2d 277 (Cal. Ct. App. 2002).
50. 148 F. Supp. 2d 1326 (S.D. Fla. 2001).
51. U.S. Const. art. I, § 8, cl. 8.
52. *See* 1 J. Thomas McCarthy, McCarthy on Trademarks and Unfair Competition chs. 1, 2 (5th ed. 2021) [hereinafter McCarthy on Trademarks].
53. *Id.* § 2:2.
54. 15 U.S.C. § 1125(a).

most civil remedies for infringement are available in such an action.[55] This separates the United States from most other countries, which grant a trademark to the first party that files for it. Therefore, an early eye to overseas trademark registrations is often warranted.

The use of a registration notice (®) with a registered trademark may help ensure the right to recover profits and damages from an infringer in limited circumstances; however, it is not a condition of trademark protection.[56]

A Word about Domain Names

Often, the first step in trademark clearance is not a trademark search but rather a check to see whether the desired domain name is available to correspond with the trademark. Domain names are often lumped together with trademarks, but they are not actually a form of intellectual property. A domain name must be registered, through a registrar accredited by the Internet Corporation for Assigned Names and Numbers (ICANN), on a first-come, first-served basis.

If a person registers, traffics in, or uses a domain name that is identical or confusingly similar to a trademark, having a bad-faith intent to profit from the goodwill associated with the mark, the trademark owner may have a cause of action for cybersquatting under 15 U.S.C. § 1125(d)(1)(C). The court may order cancellation or transfer of the domain name.[57] The remedies on such a claim are the same as for trademark infringement.[58] Alternatively, a trademark owner may take advantage of the Uniform Dispute Resolution Procedure (UDRP) and pursue a relatively low-cost online arbitration to obtain the transfer of a domain name that was registered and used in bad faith.

State Law, Federal Law

Trademarks are governed by both federal and state laws. A trademark may be registered both with a state and with the USPTO. The federal Lanham Act[59] does not generally preempt state trademark statutes or common law.[60] Nevertheless, the Lanham Act is the basis for the vast majority of court decisions in the field.

---

55. *Id.* § 1117.
56. *Id.* § 1111.
57. *Id.* § 1125(d).
58. *Id.* § 1117(a).
59. *Id.* § 1051 *et seq.*
60. 3 McCarthy on Trademarks, *supra* note 52, § 22.2.

A state trademark registration has less legal force than a federal registration, and it has effect only within the relevant state. Almost all state trademark statutes are based on the Model State Trademark Act, the most recent version of which merely provides that the registration is admissible "as competent and sufficient proof of the registration of the mark."[61] Moreover, in light of recent court decisions that broadly interpret the Lanham Act's "use in commerce" requirement for registration, even hyperlocal businesses usually have the option to file their trademarks in the USPTO.[62]

However, there are still reasons to register trademarks at the state level. First, a state registration is usually a precondition for stating an infringement claim under a state trademark statute.[63] Also, if a mark does not qualify for federal registration, it may nevertheless be registrable in state offices that do not perform true examination of applications. For the same reason, state registration comes at a fraction of the cost of federal registration.

Due to the anomalous legal landscape for cannabis products in the United States, state trademark registrations are a crucial tool for businesses selling products that violate the Controlled Substances Act or the Federal Food, Drug, and Cosmetic Act (FDCA). These products include foods and beverages containing cannabis, even cannabidiol (CBD) from legal hemp, as well as drug paraphernalia.

The courts and the USPTO's Trademark Trial and Appeal Board (TTAB) have repeatedly interpreted the Lanham Act to require "lawful use" of a trademark as a condition for federal trademark protection.[64] Further, because of the "lawful use" requirement, at least one federal court has refused to recognize even common-law trademark rights in relation to these products.[65] Therefore, a state court may be the only effective forum for a business in a trademark action.

---

61. *Id.* § 22.5.
62. *See* Christian Faith Fellowship Church v. adidas AG, 841 F.3d 986 (Fed. Cir. 2016).
63. *See, e.g.*, 3 McCarthy on Trademarks, *supra* note 52, § 22:9.25 (text of Model State Trademark Bill—2007 version).
64. *E.g.*, *In re* Stanley Bros. Soc. Enters., LLC, 2020 U.S.P.Q.2d (BNA) 10658 (T.T.A.B. 2020) (hemp oil CBD extracts sold as component of dietary supplements constitute a violation of the FDCA and do not support lawful use of the mark in commerce).
65. *See* Kiva Health Brands LLC v. Kiva Brands Inc., 439 F. Supp. 3d 1185 (N.D. Cal. 2019) (seller of cannabis-infused chocolates could not rely on prior use of trademark as a defense to infringement claim because seller's products were illegal under federal law).

The Trademark Definition

A "trademark" is any word, name, symbol, or device, or any combination thereof, that is used to identify and distinguish the owner's goods from those manufactured or sold by others and to indicate the source of the goods, even if that source is unknown.[66] A service mark is defined the same way, but for services instead of goods.[67] Most use the term "trademark" for both trademarks and service marks.

In order to qualify as a trademark, a word, symbol, or device must be used to identify and distinguish the owner's goods or services (i.e., to identify a brand). Trademark protection may extend not only to a name or logo, but also to trade dress in the form of labeling, packaging, product configuration, and essentially anything that can be perceived by the senses, including a sound or a scent.

Any of the foregoing *may* be trademarks in the right circumstances. However, trademark protection depends upon whether the thing is capable of acting as a brand identifier for the particular product or service and whether it is actually being used as a brand identifier. A number of factors must be considered, starting with how the purported trademark is being (or will be) used. For example:

- GOT MILK? is a trademark, but not when it is used within a sentence—"Have you got milk?"
- "This Side Up" is not a trademark when used to instruct handlers of a shipping carton. However, it could be a trademark when used as a creative brand name on a wine label.
- A video does not generally function as a trademark. However, if it appears at the beginning of a motion picture, like the roaring MGM lion, a video may be a trademark.

Similarly, a company name *used as such* is not a trademark. "XYZ Company" is not a trademark when used in the phrase "Produced by XYZ Company," but it may be a trademark when it appears as a brand name on a product label.

A trademark is properly thought of as an adjective that modifies the noun identifying the goods or services, as in "COCA-COLA® soft drink" or "COCA-COLA® [brand] soft drink." In contrast, a company name is a noun that may identify the source of goods or services, rather than identifying the brand of goods or services that the company is selling.

---

66. 15 U.S.C. § 1127.
67. *Id.*

However, if a company name or trade name is also used as a trademark, it may be the company's most valuable mark.

Trademark Spectrum of Distinctiveness

The existence and scope of a trademark depends next on where it falls on the "spectrum of distinctiveness."[68] The "distinctiveness" of a trademark is a measure of its ability to identify and distinguish the goods or services of the trademark owner from those of others. At one end of the spectrum is a generic term, which is the name for the class of product or service itself.[69] A generic term lacks any distinctiveness and therefore can never be a trademark. For example, the term "apple" is generic in connection with apples and cannot be appropriated by any one fruit company.

Similarly, "WA 38" is the generic term for the COSMIC CRISP® apple variety. No apple company can own WA 38, but COSMIC CRISP is a registered trademark. Indeed, a new invention, such as a new fruit variety or a new type of baked good, must be given a generic name *in addition to* its brand name. Otherwise, the purported brand name may be used by others as the generic term for the invention regardless of source. Unless a product is patent-protected, generic terms must be available for competitors to use in identifying their like products.

Moving along the spectrum, if a proposed trademark is "merely descriptive" of the goods or services, rather than generic, then it lacks distinctiveness and is ineligible for protection. However, a descriptive trademark may eventually acquire distinctiveness—also known as secondary meaning—and become eligible for protection over time.

A descriptive mark is one that directly imparts information about a quality or characteristic of the product or service. One example is the recently registered TOASTER FLAPJACK®, for frozen ready-to-eat pancakes. The distinction between a generic term and a descriptive mark can be extremely difficult to draw, and court decisions are not always consistent.[70]

Before it acquires distinctiveness, a merely descriptive mark may be federally registered on the U.S. Supplemental Register. This registration will not be evidence of exclusive rights, but the USPTO may cite it as

---

68. *See* 2 McCarthy on Trademarks, *supra* note 52, § 11.2.
69. U.S. Patent & Trademark Office v. Booking.com B.V., 140 S. Ct. 2298 (2020).
70. 2 McCarthy on Trademarks, *supra* note 52, § 11:2; *see, e.g.*, *Booking.com B.V.*, 140 S. Ct. at 2307–08 (though BOOKING is a generic term in relation to online hotel reservation services, BOOKING.COM is a merely descriptive trademark).

a bar to later-filed confusingly similar marks. A Supplemental Register registration also enables use of the ® symbol with the trademark.

The "merely descriptive" category encompasses several other types of trademarks that are not deemed to be inherently distinctive. These include, among other things, marks that are primarily merely a surname and marks that are primarily geographically descriptive or misdescriptive.[71] The latter category involves a multipart test; not all geographic terms are primarily geographically descriptive.[72]

On a related note, when dealing with certain types of food and beverage products, such as alcoholic beverages, cheese, meats, and produce, it is important to consult the various state, national, foreign, and international rules and standards that may deem a geographic term to be a protected geographic indication limited to use on products from a certain region (e.g., Vidalia for onions) or a generic term (e.g., cheddar for cheese).[73] These rules vary from country to country. A term that is permitted for use in the United States may not be permitted abroad.

The "merely descriptive" category also generally encompasses product designs and configurations.[74] For a seller hoping to register a product configuration in the USPTO, the burden to show acquired distinctiveness is quite high. The USPTO rarely approves registration in these cases. Further, trademark protection will not be granted for a product feature that is functional in any way.[75] Nevertheless, there are some well-known food product configurations that have achieved registered mark protection, including the Pepperidge Farm Goldfish Cracker and the Hershey's Chocolate Bar.

A claimant's right of priority in a merely descriptive mark arises not upon first use, but on the date that the mark acquires distinctiveness or secondary meaning.[76] In order to acquire distinctiveness, a mark must be the subject of substantially exclusive use over time so that the relevant public comes to identify it as a brand name rather than as a mere descriptor. For registration purposes, the USPTO often presumes that a descriptive mark acquires distinctiveness after five years of substantially

---

71. 15 U.S.C. § 1052(e).
72. *See In re* Newbridge Cutlery Co., 776 F.3d 854 (Fed. Cir. 2015).
73. *See, e.g.*, USPTO, EXAMINATION GUIDE 2-20: MARKS INCLUDING GEOGRAPHIC WORDING THAT DOES NOT INDICATE GEOGRAPHIC ORIGIN OF CHEESES AND PROCESSED MEATS (2020) (USPTO-T-8).
74. Wal-Mart Stores, Inc. v. Samara Bros., Inc., 529 U.S. 205 (2000).
75. TrafFix Devices, Inc. v. Mktg. Displays, Inc., 532 U.S. 23 (2001).
76. 2 MCCARTHY ON TRADEMARKS, *supra* note 52, § 11:25.

exclusive use.[77] The more descriptive the mark, the more evidence is required to prove acquired distinctiveness. In court, and sometimes in the USPTO, it is necessary to produce evidence of successful advertising, sales, and other evidence of public exposure and recognition in order to prove acquired distinctiveness.

The remainder of the spectrum of distinctiveness encompasses marks that are deemed to be inherently distinctive: suggestive marks, arbitrary marks (such as APPLE for electronic devices), and fanciful marks (i.e., coined terms (such as GOOGLE for search engines)).

Trademark Selection and Clearance

Other considerations are whether there are any absolute bars to registration under 15 U.S.C. § 1052, or other legal or regulatory restrictions on the use of the term. For example, the proposed mark might contain or imply a regulated marketing claim. The word "organic" may be used only for products that have been certified under the U.S. Department of Agriculture's National Organic Program.[78] The word "healthy" may be used on a food product only if it meets specific U.S. Food and Drug Administration (FDA) requirements, and FDA regulations contain a host of other rules on the use of certain terms.[79] The USPTO maintains a non-exhaustive list of terms that are statutorily restricted and other statutes that may affect trademarks.[80]

A clearance search is used to determine whether a proposed mark is likely to conflict with any prior marks, as well as whether the proposed mark is likely to be registrable in the USPTO. A thorough search will include not only pending and registered marks at the USPTO, but also state trademark registrations and unregistered marks in use. A standard Internet search engine is an important tool in the process.

The Likelihood of Confusion

There are two key dimensions to analyzing trademark search results: (1) whether the proposed mark is likely to be accused of infringing a prior mark and (2) whether the proposed mark is likely to be refused registration in the USPTO on the ground of conflict with a prior pending

---

77. 15 U.S.C. § 1052(f).
78. 7 U.S.C. § 6505.
79. *See* FDA, A Food Labeling Guide: Guidance for Industry (2013), https://www.fda.gov/regulatory-information/search-fda-guidance-documents/guidance-industry-food-labeling-guide.
80. USPTO, Trademark Manual of Examining Procedure app. C (2018).

or registered mark. Either way, the essential question is whether the use of the proposed mark on the relevant goods or services is likely to cause confusion, or to cause mistake, or to deceive potential purchasers as to the source, affiliation, or sponsorship of the product or the producer.[81]

The USPTO and the courts assess the likelihood of confusion with reference to a non-exhaustive list of factors that varies from jurisdiction to jurisdiction.[82] These factors are not intended to be a rigid test, but to guide the fact finder in predicting the mind of a potential purchaser of the relevant goods or services. Any facts relevant to this mind-reading exercise may be considered.

The most important factors in determining the likelihood of confusion in any case are usually the degree of similarity of the marks and the proximity or relatedness of the goods and/or services. In most cases, the fact finder will also consider the strength of the senior mark and the marketing channels used to sell the respective goods or services. In addition, evidence of either intentional infringement or actual purchaser confusion can powerfully tilt the balance in favor of a "likelihood of confusion" finding.

The stronger the senior mark, the lesser similarity between the marks will be required. "Strength" refers to a mark's capacity to identify and distinguish the owner's goods or services. There are two kinds of trademark strength: inherent and acquired.[83] Inherent strength is a function of where the mark falls on the spectrum of distinctiveness. Acquired strength is the degree of acquired distinctiveness, which is assessed according to the same kinds of evidence used to determine whether a descriptive mark has acquired secondary meaning.

The strongest marks are those that have achieved fame and are therefore entitled to protection not only against infringement by the use of similar marks on related goods, but also against dilution by the use of similar marks on unrelated goods.[84]

---

81. 15 U.S.C. § 1052(d) (examination of trademark applications); *id.* § 1114 (infringement of registered trademarks); *id.* § 1125(a) (infringement of unregistered trademarks).
82. 4 MCCARTHY ON TRADEMARKS, *supra* note 52, §§ 24:31–24:43.
83. 2 MCCARTHY ON TRADEMARKS, *supra* note 52, § 11:2.
84. 15 U.S.C. § 1125(c).

Trademark Registration

Although registration is not a precondition for trademark protection in the United States, there are many legal and quasi-legal reasons to obtain federal registration. Among them:

- A registration is prima facie evidence of the owner's exclusive rights in the mark.[85]
- A registered trademark has a nationwide right of priority, effective on the date of filing.[86]
- Many e-commerce and social media platforms, notably Amazon, will respond to a demand for takedown of infringing material only if the complainant's trademark is registered.
- A registered trademark may be recorded with the customs service, which can then be enlisted to detain infringing imports.[87]
- The ® symbol may be used with a registered trademark.[88]

Unless a trademark application is on a foreign registration under the Paris Convention for the Protection of Industrial Property or the Protocol Relating to the Madrid Agreement concerning the International Registration of Marks (Madrid Protocol), a trademark must be used in the United States in the sale or delivery of goods or services in order to achieve federal registration. However, it is unnecessary and usually undesirable to wait for commercial use before filing a trademark application. A federal application may be filed on the basis of a bona fide intention to use the mark, in which case a statement of use must be filed before the USPTO will issue the registration.[89]

Trademark registration in the USPTO is a multi-step process. Several months after the filing, a USPTO attorney will examine it for registrability. Most often, key decisions are (1) whether the proposed mark is inherently distinctive,[90] and (2) whether the proposed mark is so similar to a previously filed or registered mark as to be likely to cause confusion, mistake, or deception.[91]

If the application passes examination, the USPTO will publish the mark for opposition. At that point, interested parties have an extendable 30-day period in which to oppose the registration. An opposition

---

85. *Id.* § 1057(b).
86. *Id.* § 1057(c).
87. 19 C.F.R. § 133.1.
88. 15 U.S.C. § 1111.
89. *Id.* § 1052(b).
90. *Id.* § 1052(e).
91. *Id.* § 1052(d).

is a litigated case before the TTAB. The Federal Rules of Civil Procedure largely apply, except that proceedings are almost always conducted entirely in writing. The TTAB's authority is entirely limited to the question of registrability.

If there is no opposition, or after an opposition is favorably resolved, the trademark application can go one of two ways. If a statement of use was filed prior to publication, the USPTO will issue the certificate of registration. Otherwise, the USPTO will issue a notice of allowance. After the notice of allowance date, the applicant will have a six-month period in which to file a statement of use. This period can be extended in six-month increments for up to three years after the notice of allowance.

Trademark Infringement Remedies

Courts have broad discretion to order remedies in Lanham Act cases. The court may order an injunction to prevent a Lanham Act violation.[92] The basic monetary remedies are actual damages and defendant's profits, plus up to three times the amount of actual damages.[93] In addition, however, "if the amount of the recovery based on profits is either inadequate or excessive the court may in its discretion enter judgment for such sum as the court shall find to be just, according to the circumstances of the case."[94] In exceptional cases, usually involving willful infringement or bad conduct in the litigation, a court may award attorneys' fees to the prevailing party.[95] Additional remedies are available in counterfeiting cases.[96]

Copyright

Companies in the food industry rarely depend on copyright as a material foundation for their businesses. Copyright protection is generally unavailable for recipes or for food items, and it does not apply to methods. Copyright might protect the particular creative way that a recipe is written (expressed), but not the mere recitation of ingredients and steps.[97] Rather, the law of trade secrets is the main tool for protecting recipes and methods.

---

92. *Id.* § 1116.
93. *Id.* § 1117(a).
94. *Id.*
95. *Id.*
96. *Id.* §§ 1116(d), 1117(b).
97. *See* 1 Melville B. Nimmer & David Nimmer, Nimmer on Copyright § 2.18[I] (2015) [hereinafter Nimmer on Copyright].

Outside of content-based industries like publishing, music, entertainment, and software, copyright issues tend to arise most often in relation to labeling, marketing, and advertising. One common scenario is the receipt of a demand letter from an agency representing a photographer whose image was used on a company website without permission. Other common copyright-related projects involve securing rights from creative talent, such as graphic designers and website developers.

Copyright is overwhelmingly governed by the federal Copyright Act,[98] which preempts state laws that effectively grant the same rights.[99]

The Copyright Definition

Copyright protection applies to "original works of authorship fixed in any tangible medium of expression, now known or later developed, from which they can be perceived, reproduced, or otherwise communicated, either directly or with the aid of a machine or device."[100] Works of authorship include literary works; musical works; dramatic works; pictorial, graphic, and sculptural works; motion pictures; and other audiovisual works, sound recordings, and architectural works, among others.[101]

Copyright will not protect a "useful article," meaning an object that has an intrinsic utilitarian function other than merely to convey the appearance of the article or to convey information.[102] A cake may embody the baker's creativity, but it cannot be copyrighted as such. The decoration on a cake, however, might be eligible.

Nor does copyright protect mere ideas, concepts, or principles. Rather, copyright protects only the *expression* of an idea, and then only if it is fixed in a tangible medium of expression such as an electronic file or a piece of paper.

The use of a copyright notice bars the mitigation of damages based on innocent infringement, but it is not a condition of copyright protection.[103]

Copyright Registration

As with trademark, a registration is not required in order for a copyright to come into existence. Rather, the right arises upon creation of the work. However, in contrast to trademark, one cannot file a suit for copyright infringement until a registration has been granted by the Copyright

---

98. 17 U.S.C. § 101 *et seq.*
99. 1 Nimmer on Copyright, *supra* note 97, § 1.01.
100. 17 U.S.C. § 102.
101. *Id.*
102. *Id.* § 101.
103. *Id.* § 401.

Office.[104] In the usual case, registration takes several months from filing; however, the Copyright Office will expedite its examination upon payment of a surcharge.

Further, if a copyright is registered before it is infringed, the copyright owner may be entitled to recover statutory damages of up to $30,000 per infringement ($150,000 for willful infringement) or prevailing party attorneys' fees.[105] Since actual damages can be difficult to prove, early registration provides significant settlement leverage.

Once the authorship of the work has been determined, copyright registration is usually a fairly simple project. It requires only the completion of a two-page form, the payment of a small fee, and a deposit of the work with the Copyright Office. Trade secret information may be redacted.

Copyright Ownership

Often, companies may wrongly assume that they own the copyright if they paid for the creation of a work. In fact, the analytical starting point is that the individual who creates a work of authorship is the owner of the copyright.[106] The "work for hire" doctrine is an exception to this rule, but it applies only in two limited circumstances. First, if a company employee creates a work within the scope of his or her employment, then the company is deemed to be the author of that work in the absence of an agreement to the contrary.[107]

Second, the work product of an independent contractor may be deemed a work for hire, but only if (1) there is a written agreement saying so, and (2) the work is one of nine qualifying types.[108] A typical contract for the creation of copyrighted work may remain silent as to copyright ownership, in which case the contractor will own the copyright, or it may expressly leave ownership with the contractor. Alternatively, the agreement may recite that the developed work will be a work for hire and that rights in the work are expressly assigned to the commissioning party.

Although such a contract may be referred to as a "work for hire agreement," it usually operates as an assignment rather than actually creating a work for hire, because most works are not one of the nine statutory types. In drafting a contract containing a copyright assignment, care should be taken to specify either a present assignment effective upon creation of the work, an assignment made conditional on some later act or event, such

---

104. Fourth Estate Pub. Benefit Corp. v. Wall-Street.com, 139 S. Ct. 881 (2019).
105. 17 U.S.C. §§ 411(a), 412, 504.
106. *Id.* § 201(a).
107. *Id.* §§ 101, 201(b).
108. *Id.* § 101.

as the payment of fees, or an obligation for the contractor to assign copyright at some later point.

A marketer can minimize exposure to copyright claims by maintaining internal procedures to ensure that the authorship and ownership of each copyrightable work it uses is accounted for and documented. Relevant documentation may include, for example, online "click to agree" licenses for the download of paid or royalty-free content, signed "work for hire" and assignment agreements, and the names of all employees who contributed to the development of the work.

Copyright Fair Use

The copyright fair use defense is often unhelpful in the context of advertising or marketing, though it is otherwise extremely important. The Copyright Act lists four non-exclusive factors for courts to consider in determining whether a particular use of a copyrighted work is fair and therefore not an infringement:

1. The purpose and character of the use, including whether such use is of a commercial nature or is for nonprofit educational purposes, with "criticism, comment, news reporting, teaching, scholarship or research" most appropriate for a finding of fair use
2. The nature of the copyrighted work
3. The amount and substantiality of the portion of the copyrighted work used
4. The effect of the use on the potential market for or value of the copyrighted work[109]

In line with the first factor, the commercial use of a work in advertising or marketing might be found fair, but it is unlikely.[110] The prospects for a fair use defense may be greater if the use is "transformative" in that it adds some new expression, meaning, or messages, with a different purpose or character than the original work.[111] Under the last factor, if the defendant could have obtained a license for use, it is relatively unlikely that fair use will be found.

Copyright Infringement and Remedies

Copyright protects against copying. In order to prove copyright infringement, the owner must demonstrate that the work was actually copied.

---

109. *Id.* § 107.
110. 4 NIMMER ON COPYRIGHT, *supra* note 97, § 13.05[1][c].
111. Campbell v. Acuff-Rose Music, Inc., 510 U.S. 569 (1994).

This can be proven by showing that the defendant had access to the copyright owner's work and that there is substantial similarity between the two works.[112]

A court may grant an injunction to prevent or restrain infringement of a copyright, or order that any infringing goods be impounded.[113] Monetary damages may include any actual damages and any profits of the infringer that are not accounted for in computing the actual damages.[114] If the copyright was registered prior to the infringement, the plaintiff may instead elect to recover statutory damages, the amount of which will be determined in the court's discretion.[115]

Exhaustion and First-Sale Doctrines

Once an intellectual property rights owner sells a unit of product, certain of its intellectual property rights are "exhausted" as to that unit.[116] Under each of the three intellectual property regimes—patent, trademark, and copyright—there is an extensive body of law providing essentially that an owner of intellectual property cannot complain about the unauthorized resale of a unit of product that was initially the subject of an authorized sale by the owner. The Copyright Act codifies the rule, providing that the owner of a particular copy that was "lawfully made under this title" may sell or otherwise dispose of that copy.[117] With respect to patents and trademarks, the doctrine is a creature of case law.[118]

*****

Intellectual property is increasingly important in every industry, and the food industry is no exception. Companies in the food industry often find it particularly important to take extra care to protect their trade secrets involving products and processes. Particularly if a company is consumer-facing, its attorneys should not underestimate the complexity of trademark law. If nothing else, every proposed brand name and product name should be searched before it is used. Finally, it is best to devise some basic protocols to avoid inadvertently committing copyright infringement in the course of advertising and in a marketing context.

---

112. 4 NIMMER ON COPYRIGHT, *supra* note 97, § 13.01.
113. 17 U.S.C. §§ 502, 503.
114. *Id.* § 504.
115. *Id.* § 504(c).
116. 4 MCCARTHY ON TRADEMARKS, *supra* note 52, § 25:41.
117. 17 U.S.C. § 109.
118. 5 CHISUM ON PATENTS, *supra* note 2, § 16.03; 4 MCCARTHY ON TRADEMARKS, *supra* note 52, § 25:41.

Table of Cases

Acme Steak Co., Inc. v. Great Lakes Mech. Co., No. 98-C.A.-146, 2000 Ohio App. LEXIS 4578 (Ohio Ct. App. Sept. 29, 2000), 85 n.77

Alice Corp. Pty v. CLS Bank Int'l, 573 U.S. 208 (2014), 161 n.3

Anderson Seafoods, Inc., United States v., 622 F.2d 157 (5th Cir. 1980), 11 n.20

Ang v. Whitewave Foods Co., No. 13-CV-1953, 2013 WL 6492353 (N.D. Cal. Dec. 10, 2013), 41 n.18, 47 n.50

AT&T Mobility LLC v. Concepcion, 563 U.S. 333 (2011), 25, 41

Barber v. Nestle USA, Inc., 154 F. Supp. 3d 954 (C.D. Cal. 2015), 51 n.73

Barilla Am., Inc. v. Wright, No.4-02-CV-90267, 2002 WL 31165069 (S.D. Iowa July 5, 2002), 169 n.45

Barreto v. Westbrae Natural, Inc., No. 19-cv-9677 (PKC), 2021 WL 76331 (S.D.N.Y. Jan. 7, 2021), 52 n.85

Beardsall v. CVS Pharmacy, Inc., 953 F.3d 969 (7th Cir. 2020), 44 n.34, 46 n.41

Becerra v. Dr Pepper/Seven Up, Inc., 945 F.3d 1225 (9th Cir. 2019), 32 n.29, 46 n.43, 50

Belfiore v. Procter & Gamble Co., 311 F.R.D. 29 (E.D.N.Y. 2015), 48 n.61

Bell Atl. Corp. v. Twombly, 550 U.S. 544 (2007), 32 n.28

Bell v. Nash-Finch Co., No. 97-2191, 1999 U.S. App. LEXIS 6021 (4th Cir. Apr. 2, 1999), 89 n.123

Bell v. Publix Super Markets, Inc., 982 F.3d 468 (7th Cir. 2020), 49–50 n.67

Bimbo Bakeries USA, Inc. v. Sycamore, 372 F. Supp. 3d 1291 (D. Utah 2019), 166 n.34

Bimbo Bakeries USA, Inc. v. Sycamore, No. 2:13-cv-00749-DN-DBP, 2018 U.S. Dist. LEXIS 54556 (D. Utah Mar. 29, 2018), 166

Blue Buffalo Co. v. Nestle Purina Petcare Co., No. 4:15 CV 384 RWS, 2015 WL 3645262 (E.D. Mo. June 10, 2015), 50 n.72

Blue Ribbon Smoked Fish, Inc., United States v., 179 F. Supp. 2d 30 (E.D.N.Y. 2001), 67 n.126

Booking.com B.V., U.S. Patent and Trademark Office v. Booking.com B.V., 140 S.Ct. 2298 (2020).

Bowman v. Monsanto Co., 569 U.S. 278 (2013), 162 n.9

Bradley v. Cooper Tire & Rubber Co., No. 4:03cv00094-DPJ-JCS, 2007 U.S. Dist. LEXIS 95967 (S.D. Miss. Aug. 3, 2007), 90 n.125

Brady v. Bayer Corp., 26 Cal. App. 5th 1156 (Cal. Ct. App. 2018), 43–44

Brod v. Sioux Honey Ass'n, Coop., 927 F. Supp. 2d 811 (N.D. Cal. 2013),
    aff'd, 609 F. App'x 415 (9th Cir. 2015), 46 n.43
Bryan v. Emerson Elec. Co., 856 F.2d 192 (6th Cir. 1988), 90 n.125
Buckman v. Bombadier Corp., 893 F. Supp. 547 (E.D.N.C. 1995), 90 n.125

Campbell v. Acuff-Rose Music, Inc., 510 U.S. 569 (1994), 182 n.111
Cardinale v. Quorn Foods, Inc., No. X05CV096002022S, 2011 Conn. Super.
    LEXIS 1262 (May 19, 2011), 66 n.122
Carrea v. Dreyer's Grand Ice Cream, Inc., 475 F. App'x 113 (9th Cir. 2012),
    32 n.30
Chamberlan v. Ford Motor Co., 402 F.3d 952 (9th Cir. 2005), 30 n.21
Chase v. Gen. Motors Corp., 856 F.2d 17 (4th Cir. 1988), 90 n.125
Chen v. Dunkin' Brands, 954 F.3d 492 (2d Cir. 2020), 50 nn.70–71
Cheslow v. Ghirardelli Chocolate Co., 445 F. Supp. 3d 8 (N.D. Cal. 2020), 48,
    52 n.83
Cheslow v. Ghirardelli Chocolate Co., No. 19-CV-07467-PJH, 2020 WL
    4039365 (N.D. Cal. July 17, 2020), 51 n.75
Christian Faith Fellowship Church v. adidas AG, 841 F.3d 986 (Fed. Cir. 2016),
    172 n.62
Chung's Prods. LLP, United States v., 941 F. Supp. 2d 770 (S.D. Tex. 2013),
    62 n.71
Columbia & Puget Sound R.R. Co. v. Hawthorne, 144 U.S. 202 (1892),
    89 nn.121–122
Cosgrove v. Blue Diamond Growers, No. 19 Civ. 8993 (VM), 2020 WL
    7211218 (S.D.N.Y. Dec. 7, 2020), 52 n.85
Cruz v. Anheuser-Busch Cos., LLC, 682 F. App'x 583 (9th Cir. 2017), 32 n.30

Daley v. Weinberger, 400 F. Supp. 1288 (E.D.N.Y. 1975), 59 n.49
Daniel v. Mondelez Int'l, Inc., 287 F. Supp. 3d 177 (E.D.N.Y. 2018), 33 n.33,
    46 n.46
Del Monte Fresh Produce Co. v. Dole Food Co., 148 F. Supp. 2d 1326 (S.D.
    Fla. 2001), 170
Delgado v. Ocwen Loan Servicing, LLC, No. 13-CV-4427, 2014 WL 4773991
    (E.D.N.Y. Sept. 24, 2014), 49 n.63
Dimare Fresh, Inc. v. United States, 808 F.3d 1301 (Fed. Cir. 2015), 77 n.29
Dumont v. Reily Foods Co., 934 F.3d 35 (1st Cir. 2019), 47 n.47

Ebner v. Fresh, Inc., 838 F.3d 958 (9th Cir. 2016), 32 n.29, 44 n.36, 46 n.41,
    47 n.52, 48
Edward C. v. City of Albuquerque, 241 P.3d 1086 (N.M. 2010), 28–29 n.19

Federal Trade Commission v. R.F. Keppel & Bro., 291 U.S. 304 (1934), 38 n.7
Figy v. Frito-Lay N. Am., Inc., 67 F. Supp. 3d 1075 (N.D. Cal. 2014), 48 n.60
Fink v. Time Warner Cable, 714 F.3d 739 (2d Cir. 2013), 43 nn.26–27, 48 n.61

Fourth Estate Pub. Benefit Corp. v. Wall-Street.com, 139 S. Ct. 881 (2019), 181 n.104
Freeman v. Time, Inc., 68 F.3d 285 (9th Cir. 1995), 46 n.45, 49

Galanis v. Starbucks Corp., No. 16 C 4705, 2016 WL 6037962 (N.D. Ill. Oct. 14, 2016), 47 n.48
Gel Spice Co., United States v., 601 F. Supp. 1214 (E.D.N.Y. 1985), 61 n.65
Gibbons v. Bristol-Myers Squibb Co., 919 F.3d 699 (2d Cir. 2019), 46 n.44
Giglio v. Saab-Scania of Am., Inc., No. 90-2465, 1992 U.S. Dist. LEXIS 17026 (E.D. La. Oct. 30, 1992), 90 nn.125–126
Glen Holly Entm't, Inc. v. Tektronix Inc., 343 F.3d 1000 (9th Cir. 2003), 51 n.79
Greenfield v. Yucatan Foods, L.P., 18 F. Supp. 3d 1371 (S.D. Fla. 2014), 34 n.36

Halo Elecs., Inc. v. Pulse Elecs., Inc., 136 S. Ct. 1923 (2016), 164 n.22
Hammes v. Yamaha Motor Corp. U.S.A., Inc., No. 03-6456 (MJD/JSM), 2006 U.S. Dist. LEXIS 26526 (D. Minn. May 4, 2006), 90 n.125
Haskell v. Time, Inc., 857 F. Supp. 1392 (E.D. Cal. 1994), 46 n.45
Heinz v. Kirchner, 63 F.T.C. 1282 (1963), 40 n.13
Horizon Organic Milk Plus DHA Omega-3 Mktg. & Sales Practice Litig., *In re*, 955 F. Supp. 2d 1311 (S.D. Fla. 2013), 34 n.37
Hughes v. Ester C Co., 330 F. Supp. 3d 862 (E.D.N.Y. 2018), 46 n.40
Hughes v. Ester C Co., 930 F. Supp. 2d 439 (E.D.N.Y. 2013), 48–49

In re. See name of party

Jessani v. Monini N. Am., Inc., 744 F. App'x 18 (2d Cir. 2018), 46 n.41
Jones v. ConAgra Foods, Inc., No. C 12-01633 CRB, 2014 WL 2702726 (N.D. Cal. June 13, 2014), 42 n.23

Kiva Health Brands LLC v. Kiva Brands Inc., 439 F. Supp. 3d 1185 (N.D. Cal. 2019), 172 n.65
Koenig v. Boulder Brands, Inc., 995 F. Supp. 2d 274 (S.D.N.Y. 2014), 43 n.27, 49 n.63
Kommer v. Bayer Consumer Health, 252 F. Supp. 3d 304 (S.D.N.Y. 2017), 49 n.63

Lanard Toys Ltd. v. Dolgencorp LLC, 958 F.3d 1337 (Fed. Cir. 2020), 163 n.17
Landry v. Adam, 282 So. 2d 590 (La. Ct. App. 1973), 90 n.125, 91 n.127
Lavie v. Procter & Gamble Co., 105 Cal. App. 4th 496 (Cal. Ct. App. 2003), 43 n.25, 44 n.36
Lexington Mill & Elevator Co., United States v., 232 U.S. 399 (1914), 1–2, 10 n.18

Mantikas v. Kellogg Co., 910 F.3d 633 (2d Cir. 2018), 49 n.67
Manuel v. Pepsi-Cola Co., No. 17 CIV. 7955(PAE), 2018 WL 2269247 (S.D.N.Y. May 17, 2018), *aff'd*, 763 F. App'x 108 (2d Cir. 2019), 43 n.28
Mazza v. Am. Honda Motor Co., 666 F.3d 581 (9th Cir. 2012), 42 n.24
McDonald's French Fries Litig., *In re,* 503 F. Supp. 2d 953 (N.D. Ill. 2007), 66 n.121
Miyoko's Kitchen v. Ross, No. 3:20CV00893 (N.D. Cal. Aug. 21, 2020), 49 n.65
Moore v. Mars Petcare US, Inc., 966 F.3d 1007 (9th Cir. 2020), 47 n.54, 49 n.66
Moreno v. Vi-jon, Inc., No. 20CV1446 JM(BGS), 2021 WL 807683 (S.D. Cal. Mar. 3, 2021), 45 n.38, 50 n.68
Morse v. Minneapolis & St. Louis Ry., 16 N.W. 358 (1883), 89 nn.121–122

Nat'l Labor Relations Bd. v. Federbush Co., 121 F.2d 954 (2d Cir. 1941), 43 n.27
Neb. Beef, Ltd. v. Meyer Foods Holdings, LLC, No. 8:09-cv-00043, 2011 U.S. Dist. LEXIS 23284 (D. Neb. Feb. 24, 2011), 84 n.76
Newbridge Cutlery Co., *In re,* 776 F.3d 854 (Fed. Cir. 2015), 175 n.72
Nutrilab v. Schweiker, 713 F.2d 335 (7th Cir. 1983), 4 n.3
N.Y. Fish, Inc., United States v., No. 13-cv-2909, 2014 U.S. Dist. LEXIS 42716 (E.D.N.Y. Mar. 30, 2014), 67 n.126

Odom v. Fairbanks Mem'l Hosp., 999 P.2d 123 (Alaska 2000), 27 n.14

Painter v. Blue Diamond Growers, 757 F. App'x 517 (9th Cir. 2018), 32 n.30
Pan Am. World Airways v. United States, 371 U.S. 296 (1963), 38–39 n.7
Paramount Farms, Inc. v. Ventilex, B.V., No. CV f 08-1027 LJO SKO, 2011 U.S. Dist. LEXIS 1902 (C.D. Cal. Jan. 3, 2011), 69 n.136
Parks v. Ainsworth Pet Nutrition, LLC, 377 F. Supp. 3d 241 (S.D.N.Y. 2019), 31 n.24
PepsiCo, Inc. v. Redmond, 54 F.3d 1262 (7th Cir. 1995), 169–170
Pichardo v. Only What You Need, Inc., No. 20-CV-493 (VEC), 2020 WL 6323775 (S.D.N.Y. Oct. 27, 2020), 52 nn.85–86
POM Wonderful LLC v. Coca-Cola Co., 573 U.S. 102 (2014), 32
Procter & Gamble Co. v. Nabisco Brands, Inc., 711 F. Supp. 759 (D. Del. 1989), 163 n.14
Propulsid Prods. Liab. Litig., *In re,* No. 1355, 2003 U.S. Dist. LEXIS 3824 (E.D. La. Mar. 10, 2003), 90 nn.125–126

R. v. Smith (Morgan), 1 A.C. 146 (2000), 38 n.3
Ries v. Ariz. Beverages USA LLC, No. 10-01139 RS, 2013 WL 1287416 (N.D. Cal. Mar. 28, 2013), 44 n.34
Rombach v. Chang, 355 F.3d 164 (2d Cir. 2004), 33 n.34

Salters v. Beam Suntory, Inc., No. 4:14CV659-RH/CAS, 2015 WL 2124939 (N.D. Fla. May 1, 2015), 41–42 nn.19–20
Schering Corp., *In re,* 118 F.T.C. 1030 (1994), 21 n.43
Solak v. Hain Celestial Group, Inc., No. 317CV0704LE- KDEP, 2018 WL 1870474 (N.D.N.Y. Apr. 17, 2018), 50 n.68, 52 n.81
Stanley Bros. Soc. Enters., LLC, *In re,* 2020 U.S.P.Q.2d (BNA) 10658 (T.T.A.B. 2020), 172 n.64
Steele v. Wegmans Food Mkts., Inc., 472 F. Supp. 3d 47 (S.D.N.Y. 2020), 52 n.85
Strumlauf v. Starbucks Corp., No. 16-CV-01306-YGR, 2018 WL 306715 (N.D. Cal. Jan. 5, 2018), 51 n.75
Suchanek v. Sturm Foods, Inc., 764 F.3d 750 (7th Cir. 2014), 46 n.40
Sugawara v. PepsiCo, Inc., No. 208CV01335-MCEJFM, 2009 WL 1439115 (E.D. Cal. May 21, 2009), 37 n.1
Syntrax Innovations, Inc., United States v., 149 F. Supp. 2d 880 (E.D. Mo. 2001), 77 n.28

Thacker, Thacker *ex rel.* v. Kroger Co., 155 F. App'x 946 (8th Cir. 2005), 89 n.120, 90 n.125
Thriftimart, Inc., United States v., 429 F.2d 1006 (9th Cir. 1970), 59 n.49
Time Warner Cable, Inc. v. DirecTV, Inc., 497 F.3d 144 (2d Cir. 2007), 51 n.78
TrafFix Devices, Inc. v. Mktg. Displays, Inc., 532 U.S. 23 (2001), 175 n.75
Tubbs v. AdvoCare Int'l, L.P., 785 F. App'x 396 (9th Cir. 2019), 46 n.42
Turek v. General Mills, Inc., 62 F.3d 423 (7th Cir. 2011), 33–34
Turtle Island Foods SPC v. Foman, 424 F. Supp. 3d 552 (E.D. Ark. 2019), 49 n.65
Twohig v. Shop-Rite Supermarkets, Inc., No. 20-CV-763 (CS), 2021 WL 518021 (S.D.N.Y. Feb. 11, 2021), 51 n.74

Union Cheese Co., United States v., 902 F. Supp. 778 (N.D. Ohio 1995), 77 n.28
United States v. See name of opposing party
Utts v. Bristol-Myers Squibb Co., 251 F. Supp. 3d 644 (S.D.N.Y. 2017), *aff'd sub nom.* Gibbons v. Bristol-Myers Squibb Co., 919 F.3d 699 (2d Cir. 2019), 46 n.44

Veal v. Citrus World, Inc., No. 2:12-CV-801-IPJ, 2013 WL 120761 (N.D. Ala. Jan. 8, 2013), 47 n.49
Vitt v. Apple Computer, Inc., 469 F. App'x 605 (9th Cir. 2012), 52 n.80
Vockie v. Gen. Motors Corp., 66 F.R.D. 57 (E.D. Pa. 1975), *aff'd,* 523 F.2d 1052 (3d Cir. 1975), 90–91 nn.125, 127

Wal-Mart Stores, Inc. v. Samara Bros., Inc., 529 U.S. 205 (2000), 175 n.74
Warner-Jenkinson Co. v. Hilton Davis Chem. Co., 520 U.S. 17 (1997), 164 n.19

Weiss v. Trader Joe's, No. 19-55841, 2021 WL 816075 (9th Cir. Mar. 3, 2021), 33 n.33, 52 n.82, 46 n.46

Weiss v. Trader Joe's Co., No. 818CV01130JLSGJS, 2018 WL 6340758 (C.D. Cal. Nov. 20, 2018), *aff'd sub nom.* Weiss v. Trader Joe's, 838 F. App'x 302 (9th Cir. 2021), 52 n.82

Welk v. Beam Suntory Imp. Co., 124 F. Supp. 3d 1039 (S.D. Cal. 2015), 47 n.51

Werbel *ex rel.* v. PepsiCo, Inc., No. C 09-04456 SBA, 2010 WL 2673860 (N.D. Cal. July 2, 2010), 37–38 n.2

Whyte v. Schlage Lock Co., 125 Cal. Rptr. 2d 277 (Cal. Ct. App. 2002), 170 n.49

Williams v. Gerber Prods. Co., 552 F.3d 934 (9th Cir. 2008), 45 n.37, 48 n.61

Workman v. Plum Inc., 141 F. Supp. 3d 1032 (N.D. Cal. 2015), 50 n.68

Wysong Corp. v. APN, Inc., 889 F.3d 267 (6th Cir. 2018), 51 n.77

Yu v. Dr Pepper Snapple Group, No. 18-cv-06664-BLF, 2020 WL 5910071 (N.D. Cal. Oct. 6, 2020), 51 n.76

Zahn, *In re,* 617 F.2d 261 (C.C.P.A. 1980), 163 n.16

Index

A
Accredited Third-Party Certification Rule, 13
Acidified and low-acid foods, 12
Additives. *See* Food additives
Adulterated food
   defense against, 134
   defined, 62–63
   economically motivated adulteration, 134
   formulation, 7–9
   fraud, 131, 132–133
   manufacture, 10–11, 13, 16
   preventive controls and countermeasures, 13, 70–71
   recalls, 74, 84
   regulations, 1–2, 7–11, 13, 16, 55, 56, 58–59
   safety, 10–11, 13, 16, 55, 56, 58–59, 60, 62–65, 70–71
   sanitation and, 28
Advertising. *See also* Food labeling
   copyrights, 180
   federal regulatory agency oversight, 5
   food litigation for false, 24, 30–31, 32, 33
   reasonable consumer standard, 31, 32–33, 37–52
   health claims, 20, 21
   nutrient content claims, 19
   regulations, 5, 18–22, 50–51
   remedies for violations, 18, 19, 21–22
   structure/function claims, 20–21
   trademarks, 160

Aflatoxin, 11
Agricultural Marketing Service (AMS), 21, 107, 109
Alaska, federal nutrition programs, 98, 99, 104–105
Alcohol and Tobacco Tax and Trade Bureau (TTB), 4–5
ALERT (Anti-Fraud Locator Using the Electronic Benefits Transfer Retailer Transactions) system, 96
Allergens. *See* Food allergens
American Rescue Plan, 156
American Samoa, federal nutrition programs, 105
AMS (Agricultural Marketing Service), 21, 107, 109
Animal drugs, 3, 7 n.9
Animal foods, 3, 13, 15
Animal welfare claims, 31
Anti-Fraud Locator Using the Electronic Benefits Transfer Retailer Transactions (ALERT) system, 96

B
Better Business Bureaus, National Advertising Division, 31–32
Bioterrorism Act (2002), 2–3
Blockchain technology, 128, 129, 137–140, 141

C
C. botulinum, 10, 12
CACFP. *See* Child and Adult Care Food Program

California
  consumer protection statutes, 26, 27, 42
  Consumers Legal Remedies Act, 26
  False Advertising Law, 50
  food labeling regulations, 50–51
  food litigation, 25–26, 27, 42, 50–51
  Safe Drinking Water and Toxic Enforcement Act (Proposition 65, 1986), 29
  Unfair Competition Law, 42, 50
Canada
  Codex Alimentarius standards, 116
  food allergens, 122
  food recalls, 124
  food safety regulations, 116, 118, 123
  packaging and food contact materials, 120–121
  Safe Food for Canadians Act, 116
Cannabis or cannabidiol, 34, 172
Centers for Disease Control and Prevention (CDC), 73–74, 146
CGMP (Current Good Manufacturing Practice), 11–12, 13, 151, 154
Child and Adult Care Food Program (CACFP)
  Dietary Guidelines, 110
  eligibility requirements, 103–104
  overview, 103–105
  purpose, 103
  reimbursement rates, 104–105
  state administration, 102, 103, 104–105
Child nutrition. *See* Infant formula; National School Lunch Program; School Breakfast Program; Summer Food Service Program
Child Nutrition Act (CNA, 1966), 97, 101, 105, 110

China
  food defense from, 134
  food safety regulations, 118
Civil penalties
  copyright infringement, 181, 183
  COVID-19 fraudulent claims, 145
  food labeling and advertising violations, 18, 21–22
  food litigation and, 23–35, 40–42. *See also* Food litigation
  food manufacture violations, 18
  food recall violations, 76–77
  food safety violations, 63–65
  reportable food incident violations, 119
  trade secret misappropriation, 166, 167
  trademark infringement, 171, 179
Class actions
  food litigation, 24, 30, 32–33, 41, 42–45, 52
  reasonable consumer standard, 32–33, 42–45, 52
  requirements for, 30
CNA (Child Nutrition Act, 1966), 97, 101, 105, 110
Codex Alimentarius standards, 116–117
Color additives, 3, 7 n.9, 56
Commodity Supplemental Food Program (CSFP), 108
Consumer Brands Association, 157
Consumer protection statutes, 26, 27–28, 30, 42–43
Controlled Substances Act (1970), 172
Copyright Act (1976), 180, 182, 183
Copyrights
  copyright notice (©), 180
  definition, 180
  fair use, 182
  limited applications, 164
  overview, 160, 179–180

ownership, 181–182
registration, 180–181
remedies for infringement, 181, 182–183
trade secrets vs., 179
COVID-19 pandemic and food system, 143–158
  federal agency perspective, 143–147
  federal nutrition programs, 93, 94, 95, 102–103, 107–108, 156–157
  food labeling, 147, 148–149, 157
  food litigation, 23
  food manufacture, 146, 149–150, 151, 154
  food recalls, 151, 155
  food safety, 129, 149–150
  food security, 144, 155–157
  food supply chain, 143, 147–150
  food work environments, 147, 150–155
  foodservice industry effects, 146–147, 149, 157
  future developments, 157–158
  health claims related to, 20 n.40, 145
  packaging, 146, 147, 151
  sanitation, 150–155
  social distancing, 152
Criminal penalties
  COVID-19 fraudulent claims, 145
  food manufacture violations, 18
  food recall violations, 76–77
  food safety violations, 63–65
  reportable food incident violations, 119
  trade secret misappropriation, 167
CSFP (Commodity Supplemental Food Program), 108
Current Good Manufacturing Practice (CGMP), 11–12, 13, 151, 154

D

Debarment, 64
Deceptive acts
  consumer protection statutes, 27, 43
  defined, 39–40
  false labeling and advertising as, 22, 38
  FTC regulation, 5, 38–40
Defective substances, 11, 28, 57, 88, 89
Defend Trade Secrets Act (DTSA, 2016), 165–167
Dietary Guidelines for Americans, 109–110
Dietary Supplement Health and Education Act (1994), 2
Dietary supplements
  COVID-19 fraudulent claims, 145
  dietary ingredient, defined, 9
  federal regulatory agency oversight, 3
  food additive exceptions, 7 n.9
  labeling, 20
  preventive control exceptions, 15
  regulations, 2, 3, 9–10, 12 n.24, 15
  structure/function claims, 20
District of Columbia, federal nutrition programs, 99, 105, 108
Domain names, 160, 171
Drugs
  animal, 3, 7 n.9
  cannabis or cannabidiol, 34, 172
  COVID-19 diagnosis, treatment, prevention, 20 n.40, 145
  food additive regulations, 8–9
  food as unapproved new, 20
  trademarks, 172
DTSA (Defend Trade Secrets Act, 2016), 165–167

E

E. coli, 10
Economic Espionage Act (1996), 165

Egg Products Inspection Act (EPIA, 1970), 4, 54, 56–57, 65, 77
Eggs, 4, 54, 56–57, 65, 77, 149
Electronic Benefit Transfer (EBT) cards, 94, 103, 106
Employee confidentiality and proprietary rights agreements, 160, 169
Employees, food work environments, 147, 150–155
Environmental practices claims, 31
Environmental Protection Agency, U.S. (EPA), 5
European Union
   Codex Alimentarius standards, 116
   food allergens, 122
   food recalls, 124
   food safety regulations, 116, 117, 118, 123
   packaging and food contact materials, 120–121
   research on issues specific to, 124–125
Exhaustion doctrine, 183

F

FAO (United Nations Food and Agriculture Organization), 116, 135
Farm Bill (2018), 94
Farmers to Families Food Box Program, 107–108
Farms, food and worker safety, 15, 16, 150
FASTER (Food Allergy Safety, Treatment, Education, and Research) Act of 2021, 122
FDA. *See* Food and Drug Administration, U.S.
FDPIR (Food Distribution Program on Indian Reservations), 99–100, 104

Federal Alcohol Administration Act (FAA Act), 4–5
Federal Food, Drug, and Cosmetic Act (FDCA, 1938)
   administration, 2, 58
   amendments, 2–3
   dietary supplements, 9
   enactment, 2
   food, defined, 3–4
   food formulation, 7, 8–9
   food labeling, 2, 22, 58
   food litigation under, 28
   food manufacture, 10–13, 16–18
   food recalls, 76–77
   food safety, 10–13, 16–18, 58, 62–63
   trademarks for products violating, 172
Federal Insecticide, Fungicide, and Rodenticide Act (1947), 5
Federal Meat Inspection Act (FMIA, 1906), 1, 4, 54, 55, 64, 77
Federal nutrition programs, 93–111
   CACFP, 102, 103–105, 110
   child nutrition, 93, 97–103, 110, 156
   COVID-19 effects, 93, 94, 95, 102–103, 107–108, 156–157
   CSFP, 108
   Dietary Guidelines for Americans, 109–110
   NSLP, 93, 97–103, 110, 156
   opportunities for involvement, 110–111
   overview, 93
   reauthorization, 97, 110–111
   SBP, 93, 97–103, 110, 156
   SNAP, 93–97, 99–100, 103, 104, 107, 156–157
   Summer Food Service Program, 103
   TEFAP, 106–108

USDA Foods Program, 108–109
WIC, 105–106, 110, 156
Federal regulatory agencies, 3–5. *See also specific agencies*
  COVID-19 perspective, 143–147
  EPA, 5
  FDA, 3–4, 5, 24, 57–60, 73–74, 144–145
  FTC, 5, 24, 145
  overview, 3–5, 24
  TTB, 4–5
  USDA, 4, 24, 54–57, 73–74, 144–145
Federal Trade Commission (FTC)
  advertising regulations, 5, 18, 21–22
  regulatory authority, 5, 24, 145
  unfair or deceptive acts regulation, 5, 38–40
Federal Trade Commission Act (FTCA, 1914), 5, 21 n.42
First Amendment, 20
First-sale doctrine, 183
FMIA (Federal Meat Inspection Act, 1906), 1, 4, 54, 55, 64, 77
FNS (Food Nutrition Service), 93–102, 105–109
FOIA (Freedom of Information Act) requests, 96–97
Food, defined, 3–4
Food additives, 2, 3, 6–9, 56
Food Additives Amendment (1958), 2, 6
Food Allergen Labeling and Consumer Protection Act (2004), 3, 66
Food allergens
  food safety and, 65–66, 70
  labeling, 3, 19, 65–66, 78–79, 121–122
  major, list of, 122
  preventive controls and countermeasures, 70
  recalls for mislabeled, 78–79, 121–122

Food Allergy Safety, Treatment, Education, and Research (FASTER) Act of 2021, 122
Food and Drug Administration, U.S. (FDA)
  CBD cases, primary jurisdiction of, 34
  Center for Biologics Evaluation and Research, 57
  Center for Food Safety and Applied Nutrition, 3, 57–58
  Center for Veterinary Medicine, 3
  Codex Alimentarius standards, 116
  COVID-19 response, 145–151, 158
  dietary supplement regulation, 9, 20
  enforcement, 3
  FDCA administration, 2, 58
  food additive regulation, 6, 7–8
  food authenticity, 132
  Food Code, 4
  food labeling regulation, 2, 4, 5, 18–22, 57–58, 123, 148–149
  food manufacture regulation, 10–13, 15–18
  food recall regulation, 17, 73–92, 124
  food safety regulation, 2, 3–4, 8, 10–13, 15–18, 53–63, 66–67, 69–71, 116, 118, 128–129, 136–137, 140
  international food law and, 116, 118–120, 122–124
  natural/all natural claims, primary jurisdiction of, 34
  Office of Regulatory Affairs, 3, 57
  Office of the Chief Counsel, 3
  Office of the Commissioner, 3
  Operation Quack Hack, 145
  registration with, 59
  regulatory authority, 3–4, 5, 24, 57–60, 73–74, 144–145
  Reportable Food Registry, 17, 119
  state agencies and, 4, 16

Food and Drug Administration, U.S. (FDA), *continued*
   technology advancement adaptations, 127–129, 136–137, 140
   trademarks and term use restrictions, 176
Food and Nutrition Act (2008), 94
Food authenticity (fraud), 131, 132–133
Food Code, 4
Food defense, 60, 131, 133–134
Food Defense Rule, 60
Food Distribution Program on Indian Reservations (FDPIR), 99–100, 104
Food formulation
   COVID-19 issues, 149
   dietary supplements, 9–10. *See also* Dietary supplements
   food additives, 2, 3, 6–9, 56
   regulations, 5–10
Food fortification, 8
Food fraud (authenticity), 131, 132–133
Food imports. *See* Imported foods
Food labeling
   allergens, 3, 19, 65–66, 78–79, 121–122
   animal welfare claims, 31
   copyrights, 180
   COVID-19 issues, 147, 148–149, 157
   environmental practices claims, 31
   federal regulatory agency oversight, 4–5, 55, 56–58
   food litigation for false, 30–31, 32, 33
      reasonable consumer standard, 31, 32–33, 37–52
   food safety and, 55, 62–66
   health claims, 20, 21
   international food law, 117, 121–122, 123
   label, defined, 18
   labeling, defined, 18
   mandatory requirements, 19
   misbranding, 62–65, 133
   natural/all natural claims, 26, 31, 34
   nutrient content claims, 19
   nutrition, 2, 19, 58
   organic, 21
   place of origin claims, 31
   principal display panel, 19
   regulations, 2–3, 4–5, 18–22, 50–51, 55, 56–58
   remedies for violations, 18, 19, 21–22
   statement of identity, 19
   structure/function claims, 20–21
   trade dress protection, 173
   voluntary requirements, 19–22
Food litigation, 23–35
   animal welfare claims, 31
   California's Proposition 65 and, 29
   case length and complexity, 27
   civil, 23–35, 40–42
   class actions, 24, 30, 32–33, 41, 42–45, 52
   common defenses, 32–34
   competitor litigation, 31–32
   consumer protection statutes and, 26, 27–28, 30, 42–43
   environmental practices claims, 31
   false labeling and advertising, 24, 30–31, 32, 33
   food recall admissibility in, 89–91
   natural/all natural claims, 26, 31, 34
   negative publicity and, 26
   negligence, 28–29
   overview, 23, 35
   place of origin claims, 31
   pleading with specificity, 33, 44
   preemption, 33–34
   pre-litigation discussions/notice, 26, 29
   primary jurisdiction, 34

product liability, 24, 28, 90
reasonable consumer standard, 31,
    32–33, 37–52
regulations and, 24, 28
rise in, 23, 24–27, 40–42
slack-fill filings, 26, 30
trends, 25–26
types of cases, 27–32
vanilla cases, 26, 31
Food manufacture
    COVID-19 era, 146, 149–150,
        151, 154
    Current Good Manufacturing
        Practice, 11–12, 13, 151, 154
    Foreign Supplier Verification
        Programs Rule, 13, 16
    HACCP requirements, 12–14
    Preventive Controls Rule for
        Animal Food, 13, 15
    Preventive Controls Rule for
        Human Food, 13–15
    Produce Safety Rule, 13, 15–16
    recalls for lapses, 17
    regulations, 10–18
    remedies for violations, 17–18
    reporting lapses, 17
    sanitation, 11–12
Food Nutrition Service (FNS),
    93–102, 105–109
Food Purchase and Distribution
    Program, 109
Food quality, 131, 132
Food recalls, 73–92
    admissibility in litigation, 89–91
    allergen mislabeling, 78–79,
        121–122
    Class I, 17, 78, 83 n.74, 85, 87
    Class II, 17, 79, 83 n.74, 85, 87
    Class III, 17, 79, 83 n.74, 85
    communications, 82, 83, 85–86
    company-initiated, 75–76, 82
    conducting, 82–92
    cost of, 124
    COVID-19 era, 151, 155

depth levels, 83
effectiveness checks, 91–92
failure or refusal to initiate, 17,
    76–77
for food manufacture lapses, 17
food safety regulations and, 55,
    61, 71
government-initiated, 76, 82
HACCP, preventive controls and,
    71, 81, 84
inspections and, 77–78, 86–88
international food law, 123–124
mandatory, 17, 74, 76, 82
mock, 82
plan or strategy for, 71, 81–83
product withdrawals, 79
public warnings and, 83
regulatory agency oversight, 73–74
remedies for violations, 76–77
reports and documentation, 82, 86
scope of, 84–85
team for, 80–81
termination, 92
types of, 75–78
voluntary, 17, 74, 75–76, 82
Food regulation, 1–22. *See also*
        specific acts and agencies
    federal agencies, 3–5, 24, 54–60,
        73–74, 143–147
    food formulation, 5–10
    food labeling and advertising,
        2–3, 4–5, 18–22, 50–51, 55,
        56–58
    food litigation and, 24, 28
    food manufacture, 10–18
    food safety, 2, 3–4, 8, 10–18,
        53–60, 62–63
    international, 116–117. *See also*
        International food law
    overview, 1–3, 22, 24
    state, 1, 4, 24. *See also* States
Food safety, 53–71
    allergens and, 65–66, 70
    COVID-19 issues, 129, 149–150

Food safety, *continued*
  culture of, fostering, 129
  defined, 132
  federal regulatory agency oversight, 3–4, 53, 54–60
  food manufacture and, 10–18, 149–150
  generally recognized as safe exception, 6–9
  inspections, 54–57, 59, 60–62, 64–65
  international food law, 123
  New Era of Smarter Food Safety, 128–129, 137, 140
  packaging and, 55, 56, 120–121
  as pillar of food system, 131–132
  preventive controls and countermeasures, 59, 66–71, 116, 117, 118, 123, 128–129, 150
  prohibited acts, 62–65
  ready-to-eat foods, 71, 115–116
  recalls and, 55, 61, 71
  regulations, 2, 3–4, 8, 10–18, 53–60, 62–63
  remedies for violations, 59, 61, 63–65
  self-policing requirements, 59
  technology advancements, 128–130, 136–137, 140
Food Safety and Inspection Service (FSIS)
  food additive regulations, 8
  food labeling regulation, 4, 55, 56–57
  food recalls, 55, 61, 74, 77–78, 79, 82–84
  inspections, 54–57, 60–61
  preventive controls and countermeasures, 67–69
  regulatory authority, 4, 54–57
  Standing Emergency Management Committee, 60

Food Safety Modernization Act (FSMA, 2011)
  enactment, 3, 59
  food authenticity, 133
  food labeling, 123
  food manufacture, 13
  food recalls, 74
  food safety, 3, 13, 59–60, 67, 116, 117, 123, 140
Food security, 131, 134–135, 144, 155–157
Food Stamp Program, 93. *See also* Supplemental Nutrition Assistance Program
Food supply chain
  COVID-19 pandemic effects, 143, 147–150
  international, 3, 13, 16, 59–60, 117–118
  preventive controls and countermeasures, 71
  regulations, 2–3, 13, 16, 59–60
  technology for traceability in, 128, 136–140
Food system. *See also specific pillars*
  five pillars of, 131–135
  food defense, 131, 133–134
  food fraud (authenticity), 131, 132–133
  food quality, 131, 132
  food safety, 131–132
  food security, 131, 134–135
  securing in COVID-19 pandemic, 143–158
Food technology. *See* Technology advancements in food industry
Food work environments, 147, 150–155
Foodborne Diseases Active Surveillance Network (FoodNet), 73–74
Foreign Supplier Verification Programs (FSVP) Rule, 13, 16, 59–60, 118

Fraud
    COVID-19 diagnosis, treatment, prevention, 145
    false labeling and advertising as, 22
    food fraud (authenticity), 131, 132–133
    food safety violations, 64, 65
    pleading with specificity, 33, 44
    SNAP, 95–96
Freedom of Information Act (FOIA) requests, 96–97
FSIS. *See* Food Safety and Inspection Service
FSMA. *See* Food Safety Modernization Act
FSVP (Foreign Supplier Verification Programs) Rule, 13, 16, 59–60, 118
FTC. *See* Federal Trade Commission
FTCA (Federal Trade Commission Act, 1914), 5, 21 n.42

G
Generally recognized as safe (GRAS) exception
    common use in food claims, 7
    food additives, 6–9
    scientific procedures claims, 7
Good Management Practice (GMP), 115
Guam, federal nutrition programs, 98, 99, 105

H
Hawaii, federal nutrition programs, 98, 99, 104–105
Hazard Analysis and Critical Control Points (HACCP)
    components of, 67–71
    corrective actions, 68, 70–71
    critical control point identification, 68
    critical limits for CCPs, 68
    defined, 67 n.123
    food manufacture, 12–14
    food recalls informed by, 81, 84
    hazard, defined, 14
    hazard analysis, 67, 69
    monitoring procedures, 68, 70
    oversight and management, 70–71
    preventive controls development, 69–70
    preventive controls expanding, 13–14
    preventive controls for food safety, 59, 66–71, 116, 117, 118, 123, 128–129, 150
    recordkeeping procedures, 68
    supply chain, 71
    verification procedures, 68–69, 71
Head Start, 100, 104
Health Canada, 116
Health claims
    COVID-19, 20 n.40, 145
    food labeling, 20, 21
    trademarks and, 176
Healthy, Hunger-Free Kids Act (HHFKA, 2010), 97, 101
HHS. *See* U.S. Department of Health and Human Services
Humane Methods of Livestock Slaughter, 55, 64

I
ICANN (Internet Corporation for Assigned Names and Numbers), 171
Imported foods
    debarment, 64
    federal regulatory agency oversight, 4
    food safety violations, 64
    importer, defined, 16
    international food law, 3, 13, 16, 59–60, 113–125

Imported foods, *continued*
  logistical chain of players for, 117–118
  regulations, 3, 4, 13, 16, 59–60, 116–117
  trademarks, 178
Indian tribal organizations, federal nutrition programs, 99–100, 104, 105, 108
Industrial sabotage, 134
Infant formula, 12 n.24, 74, 106, 134
Injunctions, as remedy, 18, 63, 77, 145, 164, 183
Inspections
  food recalls and, 77–78, 86–88
  food safety, 54–57, 59, 60–62, 64–65
  regulation requirements, 1–4, 54–57, 64–65, 77
  USDA Foods Program, 109
Intellectual property, 158–183
  copyrights, 160, 164, 179–183
  counterfeiting infringement, 133, 179
  exhaustion doctrine, 183
  first-sale doctrine, 183
  overview, 159–160, 183
  patents, 159, 161–164
  practitioner responsibilities, 160
  service marks, 173
  trade secrets, 64, 65, 159, 162, 164–170, 179
  trademarks, 159–160, 164, 170–179
Intentional Adulteration Rule, 13
International food law, 113–125. *See also* Imported foods
  allergen labeling, 121–122
  Codex Alimentarius standards, 116–117
  concept, 114
  education and training, 115–116, 119–125
  EU-specific research, 124–125
  focus areas, 120–123
  food labeling, 117, 121–122, 123
  food recalls, 123–124
  lawyers' role/work, 114–115, 117–119
  overview, 113–115, 125
  packaging and food contact materials, 120–121
  pesticide and biocide residual levels, 123
  regulatory system, 116–117
  resources on, 120
  sources of norms for, 114
  supply chain, 3, 13, 16, 59–60, 117–118
Internet Corporation for Assigned Names and Numbers (ICANN), 171
Interstate commerce, 8, 56, 58, 61, 62

L
Labeling. *See* Food labeling
Lanham Act (1946), 31–32, 171–172, 179
Listeria monocytogenes, 10, 78
Litigation. *See* Food litigation

M
Madrid Protocol (Protocol Relating to the Madrid Agreement concerning the International Registration of Marks), 178
Manufacture of food. *See* Food manufacture
Meat
  additives, 8
  recalls, 77, 84
  regulations, 1, 4, 54, 55, 64, 77
Melamine, 10, 134
Mercury, 10, 11
Miller Pesticide Amendment (1954), 2
Misbranded food
  fraud, 132–133
  labeling, 19, 20

litigation, 30
preventive controls and countermeasures, 13
recalls, 74
regulations, 1, 13, 16, 19, 20, 30, 58–59
safety, 13, 16, 58–59, 62–65
Model State Trademark Act, 172

N

National Antimicrobial Resistance Monitoring System for Enteric Bacteria (NARMS), 73–74
National Molecular Subtyping Network for Foodborne Disease Surveillance (PulseNet), 73–74
National Organic Program, 21, 176
National School Lunch Program (NSLP)
  Buy American requirement, 102
  commodity entitlements, 98–99
  COVID-19 activities, 102–103, 156
  eligibility requirements, 99–100
  meals served/nutrition standards, 101, 110
  number of participants, 99
  overview, 93, 97–103
  policy, 98
  pricing and rates, 98–99
  procurement terms, 98
  state administration, 98, 102, 103
Natural/all natural claims, 26, 31, 34
NDAs (nondisclosure agreements), 160, 168
Negligence litigation, 28–29
New Era of Smarter Food Safety, 128–129, 137, 140
New York
  consumer protection statutes, 27, 43
  food litigation, 25–26, 27, 31, 42–43, 52
  trade secrets law, 165

NLEA (Nutrition Labeling and Education Act, 1990), 2, 19, 58
Nondisclosure agreements (NDAs), 160, 168
Non-GMO foods, 21
Non-GMO Project, 21
Northern Marianas, federal nutrition programs, 105
NSLP. *See* National School Lunch Program
Nutrition
  federal programs. *See* Federal nutrition programs
  food fortification, 8
  labeling regulations, 2, 19, 58
  nutrient content claims, 18
Nutrition Labeling and Education Act (NLEA, 1990), 2, 19, 58

O

Occupational Safety and Health Act (1970), 150
Occupational Safety and Health Administration (OSHA), 150–151
Organic foods, 21, 176

P

Packaging
  COVID-19 issues, 146, 147, 151
  food contact substances, 6, 120–121
  food fraud and, 133
  food safety regulations, 55, 56, 120–121
  international food law, 120–121
  labeling. *See* Food labeling
  plastics, 121
  slack-fill litigation, 26, 30
  tampering, 133
  trade dress protection, 173
Paris Convention for the Protection of Industrial Property, 178
Patent Act (1790), 161, 163

Patents
   definition, 161–162
   design, 163–164
   effect of, 164
   limited applications, 164
   on-sale bar, 162–163
   overview, 159, 161
   plant, 161–162
   remedies for infringement, 164
   trade secrets *vs.*, 162
   utility, 159, 161–164
PDP (principal display panel), 19
Penalties. *See* Civil penalties; Criminal penalties; Remedies
Pesticides, 2, 5, 7 n.9, 123
Place of origin claims, 31
Pleadings, specificity of, in food litigation, 33, 44
Poisonous or deleterious substances
   food contact materials, 121
   food litigation, 29
   may render injurious standard, 11, 63
   ordinarily injurious standard, 10–11
   regulations, 2, 10–11, 56
   residual levels, international food law, 123
Poultry
   additives, 8
   recalls, 77, 84
   regulations, 4, 8, 54, 56, 64–65, 77
Poultry Products Inspection Act (PPIA, 1957), 4, 54, 56, 64–65, 77
Preemption, 33–34
Preventive controls and countermeasures. *See also* Hazard Analysis and Critical Control Points
   COVID-19 era, 150
   food defense, 60, 131, 133–134
   food recalls and, 71, 81, 84
   food safety, 59, 66–71, 116, 117, 118, 123, 128–129, 150
   Preventive Controls Rule for Animal Food, 13, 15
   Preventive Controls Rule for Human Food, 13–15
   Risk-Based, 116, 117, 118, 123
   technology advancements, 128–129
Primary jurisdiction doctrine, 34
Principal display panel (PDP), 19
Produce Safety Rule, 13, 15–16, 59
Product liability
   insurance, 119
   litigation, 24, 28, 90
   recalls and, 80, 90
Prohibited acts, food safety regulations, 62–65
Protocol Relating to the Madrid Agreement concerning the International Registration of Marks (Madrid Protocol), 178
Puerto Rico, federal nutrition programs, 98, 105, 108
Puffery, 51–52
PulseNet (National Molecular Subtyping Network for Foodborne Disease Surveillance), 73–74
Pure Food and Drug Act (Pure Food Act, 1906), 1–2

R
Rapid Alert System for Food and Feeds (RASFF), 124
Ready-to-eat (RTE) foods, 71, 78, 115–116
Reasonable consumer standard, 37–52
   brand names as misleading, 44
   class actions, 32–33, 42–45, 52
   common sense, 43
   consumer surveys and, 51
   context and, 48–50
   defined and applied, 42–45
   dictionary definition use, 50

doctrinal origins, 38–40
factors determining
 reasonableness, 47–52
front-back dichotomy, 44, 49–50
labeling regulation compliance,
 50–51
literal truth or falsity, 43, 47–48
motion to dismiss based on, 44–45
as objective test, 37, 38, 45, 46
ongoing debates, 45–52
overview, 32–33, 37–38, 52
puffery, 51–52
rise in food litigation and, 40–42
sophistication level, 46–47
sufficiency of proof for, 44 n.34
vanilla cases, 31
Reasonable person standard, 38
Recalls. *See* Food recalls
Regulation. *See* Food regulation
Remedies. *See also* Civil penalties;
 Criminal penalties
copyright infringement, 181,
 182–183
COVID-19 fraudulent claims, 145
food labeling and advertising
 violations, 18, 19, 21–22
food litigation, 23–35, 40–42
food manufacture violations, 17–18
food recall failure or refusal, 76–77
food safety violations, 59, 61, 63–65
patent infringement, 164
SNAP violations, 95–97
trade secret misappropriation, 166,
 167–168
trademark infringement, 171, 179
Reportable Food Registry, 17, 119
Reports, food recall, 82, 86
Restaurant Meals Program, 95
Risk-Based Preventive Control
 (RBPC) plans, 116, 117,
 118, 123. *See also* Preventive
 controls and countermeasures
RTE (ready-to-eat) foods, 71, 78,
 115–116

Richard B. Russell National School
 Lunch Act (Russell Act, 1946),
 97, 101, 102, 110
Russia, food defense from, 133–134

S
Safety. *See* Food safety
Salmonella, 10
Sanitary Transportation Rule, 13, 59,
 118
Sanitation
 adulteration of food and, 28
 COVID-19 era, 150–155
 food manufacture and, 11–12
 food recalls for lapses, 84
 food safety and, 11–12, 56, 70
 preventive controls and
  countermeasures, 70
 transportation and, 13, 59, 118
School Breakfast Program (SBP)
 Buy American requirement, 102
 commodity entitlements, 98–99
 COVID-19 activities, 102–103, 156
 eligibility requirements, 99–100
 meals served/nutrition standards,
  101, 110
 number of participants, 99
 overview, 93, 97–103
 policy, 98
 pricing and rates, 98–99
 procurement terms, 98
 state administration, 98, 102, 103
Seizures, as remedy, 18, 59, 76–77
Service marks, 173
Slack-fill litigation, 26, 30
SNAP. *See* Supplemental Nutrition
 Assistance Program
Special Milk Program, 102
Special Supplemental Nutrition
 Program for Women, Infants,
 and Children (WIC)
 administration, 105
 authorized redemption locations,
  106

Special Supplemental Nutrition Program for Women, Infants, and Children (WIC), *continued*
  COVID-19 activities, 156
  Dietary Guidelines, 110
  eligibility requirements, 105–106
  food package, 106
  overview, 105–106
States. *See also specific states*
  CACFP administration, 102, 103, 104–105
  child nutrition program administration, 98, 102, 103
  consumer protection statutes, 26, 27–28, 30, 42–43
  CSFP administration, 108
  food labeling and advertising regulation, 18, 22, 50–51
  food litigation evidence codes, 89
  food manufacture regulation, 16
  food regulation, generally, 1, 4, 24
  food safety regulation, 16, 53–54
  interstate commerce, 8, 56, 58, 61, 62
  preemption of, 33–34
  reasonable consumer standard, 43–44
  SNAP administration, 94
  TEFAP administration, 106–107
  trade secret laws, 165
  trademark laws, 171–172
  unfair or deceptive acts regulation, 38–39
  WIC administration, 105–106
Strict liability, 18, 28, 90
Structure/function claims
  COVID-19 diagnosis, treatment, prevention, 145
  food labeling, 20–21
Summer Food Service Program, 103
Supplemental Nutrition Assistance Program (SNAP)
  administrative review process, 96–97
  ALERT system, 96
  authorized retailers and entities, 94–95
  CACFP eligibility with, 104
  child nutrition program eligibility with, 99–100, 103
  COVID-19 activities, 94, 95, 103, 156–157
  Electronic Benefit Transfer cards, 94, 103
  eligible purchases, 94, 95
  fraud, 95–96
  judicial review process, 97
  overview, 93–97
  purpose, 93–94
  remedies for violations, 95–97
  TEFAP foods and, 107
Supply chain. *See* Food supply chain

T
TANF (Temporary Assistance for Needy Families), 100
Technology advancements in food industry, 127–141
  artificial intelligence, 128
  bar codes, 136
  blockchain, 128, 129, 137–140, 141
  business models and retail modernization, 129
  food safety, 128–130, 136–137, 140
  food system pillars and, 131–135
  food technology ecosystem, 135–137
  Internet of Things, 128
  overview, 127–128, 140–141
  preventive controls and countermeasures, 128–129
  QR codes, 136
  sensor technologies, 128
  smart contracts, 138–140
  tech-enabled traceability, 128, 136–140

Temporary Assistance for Needy
    Families (TANF), 100
Terrorism, 2–3, 134
The Emergency Food Assistance
    Program (TEFAP)
  administration, 106, 107
  COVID-19 response, 107–108
  eligibility requirements, 107
  foods available, 107
  operating assistance, 107
  overview, 106–108
Thrifty Food Plan, 157
Trade dress, 173
Trade secrets
  copyrights *vs.*, 179
  definition, 165–166
  economic value, 166
  employee confidentiality
      and proprietary rights
      agreements, 169
  inevitable disclosure doctrine,
      169–170
  keeping secret, 165, 166
  misappropriation, 167
  nondisclosure agreements, 168
  overview, 159, 164–165
  patents *vs.*, 162
  prohibited use, 64, 65
  remedies for misappropriation,
      166, 167–168
Trademark Trial and Appeal Board
    (TTAB), 172, 179
Trademarks
  arbitrary, 176
  company name or trade name,
      173–174
  definition, 173–174
  descriptive, 174–176, 177
  distinctiveness spectrum, 174–176,
      177, 178
  domain names, 171
  exclusive use, 175–176
  fanciful, 176
  generic terms not, 174
  goodwill and, 170, 171
  likelihood of confusion, 176–177,
      178
  limited applications, 164
  opposition to registration, 178–179
  overseas or foreign registration,
      171, 178
  overview, 159–160, 170–171
  registration, 170–172, 174–175,
      176, 178–179
  registration notice (®), 171, 175,
      178
  remedies for infringement, 171,
      179
  selection and clearance, 176, 178
  state and federal laws, 171–172
  strength of, 177
  suggestive, 176
TTAB (Trademark Trial and Appeal
    Board), 172, 179
TTB (Alcohol and Tobacco Tax and
    Trade Bureau), 4–5

U
Unfair acts
  consumer protection statutes, 27,
      42
  defined, 39
  FTC regulation, 5, 38–39
Uniform Dispute Resolution
    Procedure (UDRP), 171
Uniform Trade Secrets Act (UTSA),
    165–167
United Nations Food and Agriculture
    Organization (FAO), 116, 135
United Nations Sustainable
    Development Goals (UN
    SDGs), 134–135
U.S. Copyright Office, 180–181
U.S. Department of Agriculture
    (USDA)
  Agricultural Marketing Service,
      21, 107, 109
  Codex Alimentarius standards, 116

U.S. Department of Agriculture
(USDA), *continued*
COVID-19 response, 146, 148,
156–157, 158
Dietary Guidelines for Americans,
109–110
federal nutrition programs,
93–111, 156–157
food labeling regulation, 4, 55,
56–57, 148
Food Nutrition Service, 93–102,
105–109
food recall regulation, 55, 61,
73–74, 77–78, 79, 82–84,
87–88, 124
Food Safety and Inspection
Service, 4, 8, 54–57, 60–61,
67–69, 74, 77–78, 79, 82–84
food safety regulation, 4, 8, 53–57,
60–61, 64–69, 116, 118
international food law and, 116,
118, 120, 124
National Organic Program, 21, 176
regulatory authority, 4, 24, 54–57,
73–74, 144–145
U.S. Department of Health and
Human Services (HHS)
Dietary Guidelines for Americans,
109–110
Food and Drug Administration,
57. *See also* Food and Drug
Administration, U.S.

U.S. Department of the Treasury,
Alcohol and Tobacco Tax and
Trade Bureau, 4–5
U.S. Patent and Trademark Office
(USPTO), 159, 162–163,
171–172, 174–179
Trademark Trial and Appeal
Board, 172, 179
U.S. Supplemental Register, 174–175
USDA Foods Program, 108–109
UTSA (Uniform Trade Secrets Act),
165–167

V
Vanilla litigation, 26, 31
Vibrio vulnificus, 10
Virgin Islands, federal nutrition
programs, 98, 105

W
Water, 12, 16, 29, 52, 94, 115, 135
Websites, labeling regulations
applicable to, 18, 157
WIC. *See* Special Supplemental
Nutrition Program for Women,
Infants, and Children
Work for hire agreements, 160,
181–182
World Health Organization (WHO),
116
World Trade Organization (WTO),
116, 117